TRANSFORMING PLASTIC

FROM POLLUTION TO EVOLUTION

Albert Bates

GroundSwell Books

Summertown, Tennessee

Library of Congress Cataloging-in-Publication Data

Names: Bates, Albert K., 1947- author.
Title: Transforming plastic : from pollution to evolution / Albert Bates.
Description: Summertown, Tennessee : GroundSwell Books, [2019] | Series: Planet in crisis | Includes bibliographical references.
Identifiers: LCCN 2019007472 | ISBN 9781570673719 (pbk.)
Subjects: LCSH: Plastic scrap—Environmental aspects. | Plastics—Recycling. | Plastics—Technological innovations.
Classification: LCC TD798 .B37 2019 | DDC 363.738—dc23
LC record available at https://lccn.loc.gov/2019007472

We chose to print this title on sustainably harvested paper stock certifi ed by the Forest Stewardship Council®, an independent auditor of responsible forestry practices. For more information, visit us.fsc.org.

Printed in the United States of America

GroundSwell Books
an imprint of Book Publishing Company
PO Box 99
Summertown, TN 38483
888-260-8458
bookpubco.com

ISBN: 978-1-57067-371-9

24 23 22 21 20 19   1 2 3 4 5 6 7 8 9

Contents

Acknowledgments

All books depend upon the contributions of many. For this book I began with the assistance of Gayla Groom, using Slack to share notes and Scrivener to aggregate our research, even as we both roamed the world, sometimes fourteen time zones apart. I was thankful to have Kindle while I was sitting in airports and train stations or riding on planes and trains. I am grateful to the authors of the many written works and interviews I consulted, but above all Susan Freinkel and Michael Tolinski who elevated the discussion above the usual do's and don'ts lists. The final edition was ported from Google Docs to Microsoft Word and cross-checked in Grammarly before going to the publisher. The costs of these tools and subscriptions are not paid for by advances or will even likely be repaid by royalties, so my gratitude goes to angels such as Ross Jackson, Geoff Oelsner, Ian Graham, Doug Guyer, and Bruce and Roslyn Moore, among many others who helped in ways large and small. In the editing process, Kathy Hill, Gayla Groom, and Jo Stepaniak turned a rough work into something more polished. Bob and Cynthia Holzapfel at Book Publishing Company provided ideas and encouragement from the outset, and when they needed to be patient with my delays, they were. Finally, I owe special thanks to Alexandre Lemille at Wizeimpact, Renate Dauvarte and team at The Ocean Cleanup, Anne-Sophie Garrigou at *The Beam* magazine, Pete Kelly at Okefenokee Glee & Perloo, Jason Deptula and Millie Kellems-Otto, Carolina Erminy, Veronica Valenzuela Gibson, Maria Martinez Ros, Sandra Thomson, and Leobardo Velazquez.

Introduction

Most people come to this discussion having been moved by an image. Perhaps it was a sea turtle with a plastic straw in its nose, or a seal pup suffocating after getting tangled in a plastic bag. Maybe it was the corpse of a pelican or whale whose innards were packed with colorful plastic knickknacks. Perhaps you read a magazine article telling you that you now have microplastics in your kidneys and liver, as we all do, and as your children will, and their children. These things horrify us, and rightly so. They may make us angry, but do they empower us to do something?

I come to this subject from a dedication to regenerative system design. Whether teaching permaculture courses, working with bright young students at Gaia University, or building communities of the future with the Global Ecovillage Network, my strategy has always been to turn the amazing energy and creativity of youth toward creating a better future. After all, they have the most to lose, and if it is going to get better, they'll be the ones to make it so.

Plastics are a challenge that they will find very difficult to deal with, and they're becoming more difficult to deal with each year.

I am an EPT: emergency planetary technician. This book is part of a series that examines each of the planetary crises we now face and offers a training course in emergency care. I believe that if many more people decide to join with me and become emergency planetary technicians, we have a decent chance of stabilizing the patient.

This book takes a look at the many problems of plastic, but it only sets the stage. Its main focus is on solutions—things you can do, things governments must do, and things smart business leaders will do to turn profits. I will take you on the journey I took as I looked at the world's current predicament and the various solutions that have been explored so far.

Now is not the best time to be dealing with this. Fifty or one hundred years ago would have been better. Some of the problems caused by plastics can't be fixed now—they will be with us forever. But now is the second-best time to start, so let's go.

CHAPTER 1

Arithmetic

If a path to the better there be, it begins with a full look at the worst.

THOMAS HARDY

The late University of Colorado mathematics professor Albert Bartlett is said to have given his famous lecture on the exponential function more than one thousand times. I show the YouTube video of that lecture at the start of my permaculture courses, and I suspect many other instructors do the same. The audience continues to grow each year since Bartlett passed. In his one-hour talk, the professor says:

> Legend has it that the game of chess was invented by a mathematician who worked for a king. The king was very pleased. He said, "I want to reward you." The mathematician said, "My needs are modest. Please take my new chessboard and on the first square, place one grain of wheat. On the next square, double the one to make two. On the next square, double the two to make four. Just keep doubling till you've doubled for every square; that will be an adequate payment." We can guess the king thought, "This foolish man. I was ready to give him a real reward; all he asked for was just a few grains of wheat."
>
> But let's see what is involved in this. We know there are eight grains on the fourth square. I can get this number, eight, by multiplying three twos together. It's 2 x 2 x 2; it's one two less than the number of the square. Now that continues in each case. So on the last square, I'd find the number of grains by multiplying sixty-three twos together.

> Now let's look at the way the totals build up. When we add one grain on the first square, the total on the board is one. We add two grains, that makes a total of three. We put on four grains, now the total

> is seven. Seven is a grain less than eight, it's a grain less than three twos multiplied together. Fifteen is a grain less than four twos multiplied together. That continues in each case, so when we're done, the total number of grains will be one grain less than the number I get multiplying sixty-four twos together. My question is, how much wheat is that?
>
> You know, would that be a nice pile here in the room? Would it fill the building? Would it cover the county to a depth of two meters? How much wheat are we talking about?
>
> The answer is, it's roughly four hundred times the 1990 worldwide harvest of wheat. That could be more wheat than humans have harvested in the entire history of the earth. You say, "How did you get such a big number?" and the answer is, it was simple. We just started with one grain, but we let the number grow steadily until it had doubled a mere sixty-three times.

"The greatest failing for the human race," Bartlett was fond of telling his students, "is its failure to understand the exponential function."

> Now there's something else that's very important: the growth in any doubling time is greater than the total of all the preceding growth. For example, when I put eight grains on the fourth square, the eight is larger than the total of seven that were already there. I put thirty-two grains on the sixth square. The thirty-two is larger than the total of thirty-one that were already there. Every time the growing quantity doubles, it takes more than all you'd used in all the proceeding growth.

At another point in the lecture, Bartlett gives the example of bacteria filling a bottle. The doubling rate is every minute, so Bartlett poses this challenge to his students:

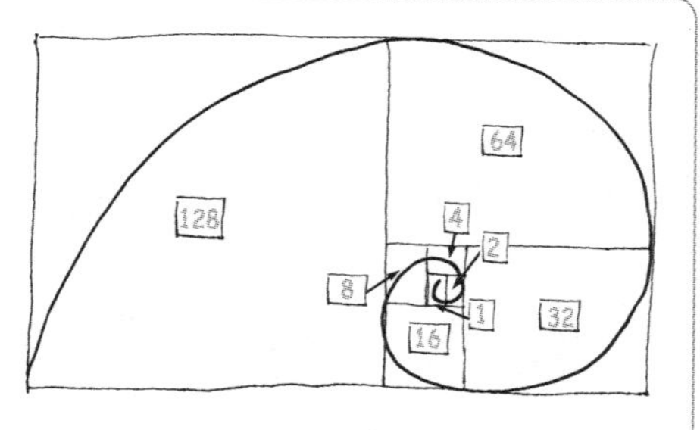

In exponential growth, each doubling provides more than all the previous doublings combined. In just seven doublings, 1 becomes 128.

> If you were an average bacterium in that bottle, at what time would you first realize you were running out of space? Well, let's just look at the last minutes in the bottle. At 12:00 noon, it's full; one minute before, it's half full; two minutes before, it's a quarter full; then an eighth; then a sixteenth. Let me ask you, at five minutes before 12:00, when the bottle is only 3 percent full and is 97 percent open space just yearning for development, how many of you would realize there's a problem?

This brings us to plastics, which are now in their fourth doubling since 1968. By any fourth doubling, the curve's trajectory is still at the bottom of the J and only beginning to bend upward. By 2030, the slope up will be much more obvious, just as it is for climate change.

Today the equivalent of five grocery bags of plastic trash piles up behind every foot of coastline on the planet, washed in from the ocean. A few years from now, that will be ten bags per foot. Huge garbage patches have formed in our oceans, created by the drift of trash from rivers into ocean currents. If we had one Great Pacific Garbage Patch twice the size of Texas in 2015, before 2030 we would have one that's four times the size of Texas and eight times the state of Texas by midcentury. If the present Garbage Patch kills one hundred thousand marine birds yearly, by midcentury it may be killing eight times that many. And that particular patch accounts for only five ten-thousandths (0.0005) of marine-mammal mortality from all plastics globally.

If each human baby born now has detectable microplastics in its blood, in ten or twenty years (depending on the doubling rate), that child's child will have twice that much, and with each generation it will double and then double again as we go through this century.

Isn't it time we asked why we design a material to last forever and then put that into objects intended to be discarded after a single use?

An honest assessment of why we, the clever bipeds, make such gargantuan faux pas in design would conclude it is because we seldom reason together. Usually we reason separately or in small groups. We run in packs. So more often than not, we consign decisions of this type to the pack that has the skills or inclination—in this case, chemical or manufacturing companies. But we need to acknowledge that these companies are in business to make a profit and make decisions that favor them most. While we can't say sustainability doesn't factor in at all, we know that *economic* sustainability over the course of one or more business cycles is what gets primary consideration. Environmental and social costs may only matter if those threaten profits.

The chemists working in well-financed research labs are only doing what they are told, which is to design something with the following attributes:

- cheap (without reference to social, environmental, disposal, or cleanup costs);

- durable (even indestructible by natural decay processes);
- lightweight (even buoyant), strong, and compact.

These design parameters apply equally to bio-based plastics, now growing at 40 percent per year (with a doubling time of fourteen months), thanks to green consumer demand. Of course, bioplastics are still plastics: still cheap, durable, and lightweight, and often just as much of an environmental problem. They may even consume as much fossil fuel to produce. Priced at a premium over their fossil-based cousins, they assuage guilt while building the bottom line.

Confronted by the next-gen market challenge to go ever greener, producers have come up with replacements for the worst plastics: the heavy-metal-based additives and coatings, halogenated flame retardants, carcinogenic styrenic petrochemicals (the ones found in polystyrene foam), endocrine-disrupting phthalate plasticizing additives, and ozone-depleting foaming agents. They have not found a substitute for the chlorine in PVC, even with corn- or cane-derived bio-PVC. There is just something about vinyl that's ... better.

A request to company executives and chemists: please stop digging us into a deeper hole. To place the burden entirely on consumers, as most "solutions" do, is unfair. In the next few years, we need new products designed to degrade under natural conditions. At a minimum, packaging materials (the largest stream of plastics) should break down into harmless components in saltwater. For products that need to function in marine environments, we might even consider replacing them with whatever was used *before* plastics.

Recycling is mostly an illusion when it comes to plastic. In the United States, plastics are recovered at lower rates from municipal solid waste than all other major material types, and for a good reason. Even when uncontaminated, separating by type and form is as hard for recycling facilities as it is for consumers. Moreover, there are technical limits on the amount of recycled resin that can be used in a given product, most resins can be reused only once, in some cases the cost of recycled plastic may be higher than virgin plastic, and the range of products where recycled content is acceptable is limited.

Since the first plastic polymers were introduced, about six billion tons of plastics have been made and spread around the planet, nearly one ton for every person now living. Even if we decide to change our plastic-using ways, the damage has already been done. Whether we like it or not, our landfills will be excavated by future archaeologists, and our Plastic Age will take its place after the Bronze Age and the Iron Age in the history of human civilization.

On Hawai'i's Kamilo Beach, a new mineral called plastiglomerate has been discovered. It has yet to be found elsewhere, but it is "natural" at least in the sense that it formed the same way many other volcanic rocks of Hawai'i have. It is the aggregate of melted plastic trash mixing with sediment, basaltic lava fragments, seashells, and organic debris. There are two types: clastic and in situ. Clastic plastiglomerate has been incorporated into rocks by heat. In situ plastiglomerate is glued together by pressure. While both kinds are found on Kamilo Beach, neither was caused by lava flows, although that could be happening elsewhere. The Kamilo plastiglomerates are produced by plastics being burned in campfires or by the plastic residues in the black beach sand being baked by the sun's rays.

Plastics are a problem for our culture, whether you believe in market forces, controlled economies, or social democracy as your favored regulatory mechanism. We need something to change, and we need it to happen quickly. The important consideration now is not whether the Plastic Age can be prevented—it's too late for that—but whether it can be shortened and made friendlier, and what kind of age will follow.

CHAPTER 2

Addiction

> I sometimes think that there is a malign force loose in the universe that is the social equivalent of cancer, and it's plastic. It infiltrates everything. It's a metastasis. It gets into every single pore of productive life.
>
> **NORMAN MAILER,** *HARVARD MAGAZINE,* 1983

I am addicted to plastic. How can I freak out about dolphins drowning in plastic nets or seagulls eating lighters and condoms off the beach, when I give no second thought to picking up a plastic comb in an airport shop, even if I decline the plastic bag?

The word "plastic" comes from the Greek verb *plassein*, which means "to mold or shape." Its flexibility derives from long, bouncy chains of carbon, oxygen, and hydrogen atoms arrayed in repeating patterns that behave like a snake's skin.

Snakeskin is a good example because biology has been knitting these molecular daisy chains for hundreds of millions of years. The cellulose that makes up the cell walls in reptiles is a polymer. Before there were plastic Wellies and galoshes, there were snakeskin boots.

"Polymer" is Greek for "many parts"; any polymer is a long chain of nearly identical molecules. The proteins that code the stems and flowers of daisies and also code our muscles, skin, and bones and the long spiraling ladders of DNA that entwine the genetic destinies of daisies and bones are all polymers. Take some of these protein chains, rearrange them slightly, and their choreography or dancers will dictate specific characteristics, just as different dance arrangements do.

> Bring chlorine into that molecular conga line, and you can get polyvinyl chloride, otherwise known as vinyl; tag on fluorine, and you can wind up with that slick nonstick material Teflon.
>
> **SUSAN FREINKEL,** AUTHOR OF *PLASTIC: A TOXIC LOVE STORY*

Take just a moment and let's walk back a step. The dancing line of carbon, oxygen, nitrogen, and hydrogen was no more than air and water, rearranged. But now we throw in chlorine and fluorine and what happens? Permanence. That substance has withdrawn from the contract with nature whereby all things must return full cycle, each with its own sunset clause.

For most of history, combs were made of almost any material humans had at hand, including bone, tortoiseshell, ivory, rubber, iron, tin, gold, silver, lead, reeds, wood, glass, porcelain, paper-mâché. But in the late nineteenth century, that panoply of possibilities began to fall away with the arrival of a totally new kind of material—celluloid, the first man-made plastic. Combs were among the first and most popular objects made of celluloid. And having crossed that material Rubicon, comb makers never went back. Ever since, combs generally have been made of one kind of plastic or another.

SUSAN FREINKEL

The first artificial plastics—celluloid combs developed in 1869 by a young inventor in upstate New York—arrived at a moment of cultural transition. The turn of the twentieth century marked the birth of the consumer culture, the global switch from growing and preparing our own food and making our own clothing (excluding the aristocracy) to consuming mass market simulacra from factories. As historian Jeffrey Meikle pointed out in *American Plastic:* "By replacing materials that were hard to find or expensive to process, celluloid democratized a host of goods for an expanding consumption-oriented middle class." Or as Susan Freinkel put it, plastics "offered a means for Americans to buy their way into new stations in life."

They also offered a way for bacteria to shirk *their* stations in life.

Unintended Consequences

Celluloid combs and cellophane tape were gateway drugs. In 1907, Leo Baekeland combined cancerous formaldehyde with phenol derived from foul-smelling and nasty coal tar, and voila! His

Bakelite was a tough, slick polymer that could be precisely molded and machined into nearly anything.

Families gathered around Bakelite radios to listen to programs sponsored by the Bakelite Corporation, drove Bakelite-accessorized cars, kept in touch with Bakelite phones, washed clothes in machines with Bakelite blades, pressed out wrinkles with Bakelite-encased irons—and, of course, styled their hair with Bakelite combs.

Bakelite inspired companies such as DuPont, Dow, Standard Oil, Union Carbide, and 3M to get into the race. Discoveries followed, and mass production of plastic products commenced. But Bakelite introduced something new to nature that was largely unappreciated at the time. Once those molecules were linked into a daisy chain, they couldn't be unlinked. Microbes don't care to spend the energy required to break those tough bonds if they can find food more obliging elsewhere.

> "From the time that a man brushes his teeth in the morning with a Bakelite-handled brush until the moment when he removes his last cigarette from a Bakelite holder, extinguishes it in a Bakelite ashtray, and falls back upon a Bakelite bed, all that he touches, sees, uses will be made of this material of a thousand purposes," *Time* magazine enthused in 1924 in an issue that sported Baekeland on the cover.
>
> **SUSAN FREINKEL**

You can break a piece of Bakelite, but you can't make it into something else. It does not degrade. It never goes away. This is why you'll still find vintage Bakelite phones, frames, radios, and combs that look nearly brand new, and why today plastic debris is piling up on land and in the open ocean, in the entrails of dead whales on shorelines, and in living crustaceans on the deepest seabed of the Marianas Trench.

In nature nothing is permanent. Everything is food for someone else. Composers and decomposers coevolved in an endless dance—

a harmony and rhythm that defines life. There is birth, and there is death. But we could not accept that.

In the last half century, there have been many drastic changes to the surface of our planet, but one of the most astonishing is the ubiquity and abundance of plastic. Even if we go extinct, that plastic will persist. We have only slowly moved from thinking of this as an aesthetic problem—litter and flotsam—to grokking that the choking wildlife we are seeing is actually a threat to *us*. Dead reefs and red tides are sending warnings: destroy the marine food chain and you'll choke your own.

When consumers first considered the permanence of plastic, we thought it was a good thing. Bakelite replaced rhino horn, elephant tusk, and tortoiseshell but was even better—cheaper, tougher, wildlife-hunter safe. Remember: All of those flesh and bone things break down over time and need to be replaced. We were running out of rhinos, elephants, and tortoises.

In 1955, *Life* magazine ran the headline "Throwaway Living" below a photograph showing a family flinging plates, cups, and cutlery into the air. The items would take forty hours to clean, *Life* said, "except that no housewife need bother." What *Life* failed to mention is that all those "disposable" items would still be around forty years later, and four hundred, and four million.

Cups, nylon stockings, radios, and telephones led to the "consumer culture"—a democratization of material comfort and leisure by making more things affordable to the masses through the clever vehicle of externalizing the true costs. The translation of plastics from camera film and nylons to the beverage and food-packaging industry further evolved the nascent consumer culture into a "throwaway culture."

Throwaway Living

DISPOSABLE ITEMS CUT DOWN HOUSEHOLD CHORES

The objects flying through the air in this picture would take 40 hours to clean—except that no housewife need bother. They are all meant to be thrown away after use. Many are new; others, such as paper plates and towels, have been around a long time but are now being made more attractive.

At the bottom of the picture, to the left of a New York City Department of Sanitation trash can, are some throwaway vases and flowers, popcorn that pops in its own pan. Moving clockwise around the photograph come assorted frozen food containers, a checkered paper napkin, a disposable diaper (seriously suggested as one reason for a rise in the U.S. birth rate) and, behind it, a baby's bib. At top are throwaway water wings, foil pans, paper tablecloth, guest towels and a sectional plate. At right is an all-purpose bucket and, scattered throughout the picture, paper cups for beer and highballs. In the basket are throwaway draperies, ash trays, garbage bags, hot pads, mats and a feeding dish for dogs. At the base of the basket are two items for hunters to throw away: disposable goose and duck decoys.

CONTINUED

Creator: Peter Stackpole for *Life*

This new way of "using once and throwing away" metamorphosed into a normal feature of ordinary everyday lives, a practice that is not merely taken for granted but also nearly impossible to avoid. Forty percent of the 450 million tons of plastic produced each year is designed to be discarded after a single use, usually within minutes of purchase. Consider the 1,200 billion plastic bottles Coca-Cola produces each year or the plastic wrappers on your "garden-fresh" produce. What about Huggies, toothbrushes, or birth control packets?

The Last Straw

If I can trace my addiction to its roots, it may have started in the 1950s with the Flav-R-Straw. Flav-R-Straws were a bendy straw with hundreds of tiny flavor pellets in the bellows that could turn plain milk into chocolate or strawberry.

Flav-R-Straws were withdrawn in 1961 but not plastic straws. 500,000,000. Five with eight zeros. That is how many plastic straws go into drink cups. Not every year—every day. And that's just in the United States.

I can remember my delight as a child when a friend showed me how to take off the end of the paper wrapper and blow the wrap-

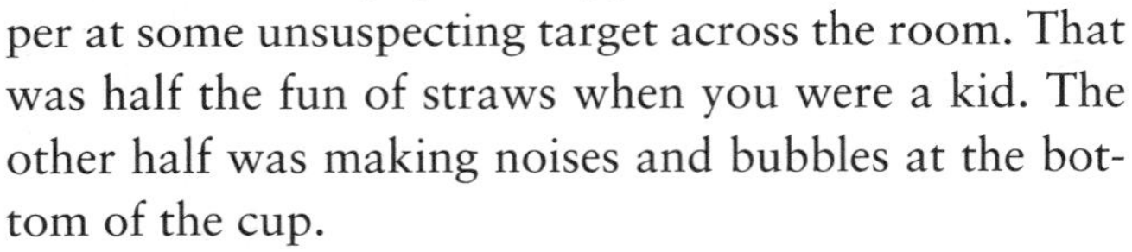

per at some unsuspecting target across the room. That was half the fun of straws when you were a kid. The other half was making noises and bubbles at the bottom of the cup.

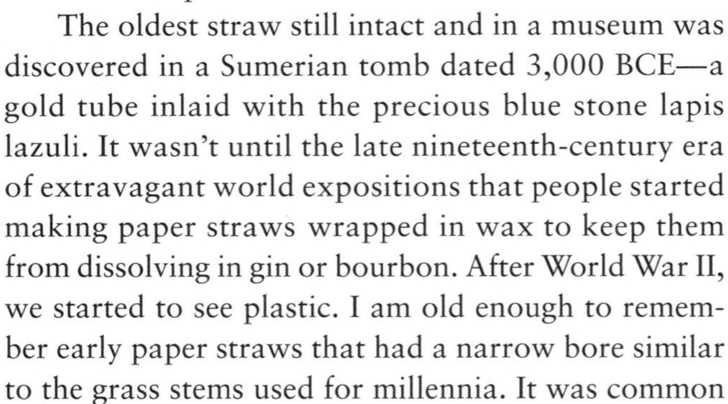

The oldest straw still intact and in a museum was discovered in a Sumerian tomb dated 3,000 BCE—a gold tube inlaid with the precious blue stone lapis lazuli. It wasn't until the late nineteenth-century era of extravagant world expositions that people started making paper straws wrapped in wax to keep them from dissolving in gin or bourbon. After World War II, we started to see plastic. I am old enough to remember early paper straws that had a narrow bore similar to the grass stems used for millennia. It was common to use two of them to reduce the effort needed to take each sip. Modern plastic straws are made with a larger bore so only one is needed for ease of drinking, but when they hand you your 64-ounce Biggie

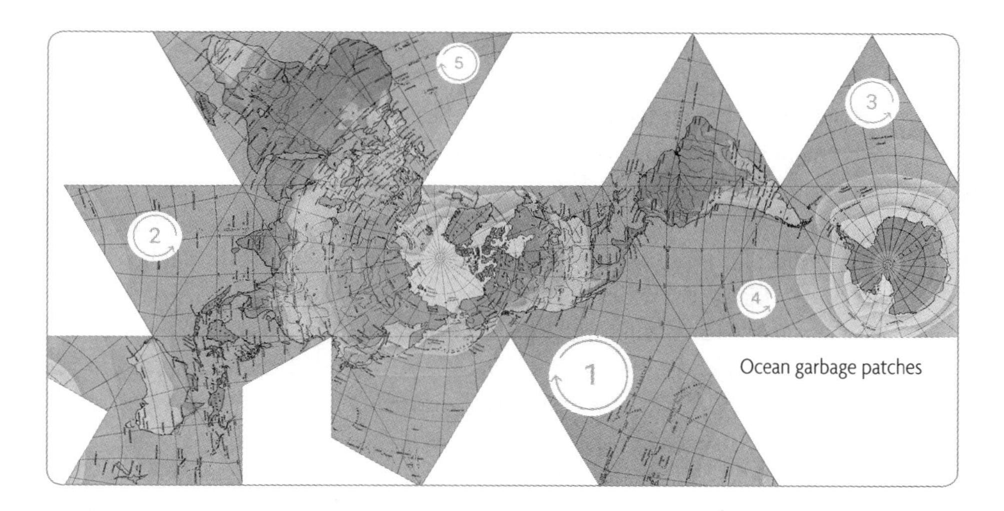

Ocean garbage patches

through the drive-up window, chances are it's got two, purely out of habit.

And it's single-use plastic.

You can complain and they will take back the straws, but when you aren't looking, those are going straight into the trash, which goes straight into a dumpster (in a plastic bag), which goes maybe to separation and maybe not, and then to either a landfill or to some watercourse that leads to the ocean and thence the gullet of seabirds or the digestive organs of fish, turtles, dolphins, and whales. Robot subs have found plastic in the stomachs of creatures thirty-six thousand feet down.

Eighty-eight to 95 percent of the plastic polluting the world's oceans pours in from just ten rivers—eight in Asia and two in Africa. These rivers account for about five trillion pounds of plastic garbage floating in the seas. It kills an estimated two hundred million marine mammals annually.

Besides the North Atlantic and North Pacific patches already discovered, there are now three spots located in the South Pacific, South Atlantic, and Indian Oceans that are subject to the same phenomenon. Actually, those are just natural concentration points. Plastic waste can be found *everywhere* in the oceans—from beaches where people go on holidays to remote uninhabited islands. Plastics have been fragmenting and accumulating in the oceans for more than fifty years, and a full recovery may never be possible.

In 2018, I had the good fortune to meet Jackie Nunez, founder of The Last Plastic Straw. She said when plastic was first applied to paper straws, straws became a gateway drug because they were so easy and ubiquitous. But that is also what makes straws a gateway solution, or "sipping point."

Nunez said, "I had my Last Plastic Straw moment in 2011 after receiving a glass of water with a plastic straw at a local beachside bar in Santa Cruz, California. I didn't ask for a straw. I had just arrived into town after traveling the Caribbean. While there, everywhere I went I saw plastic pollution. On the beaches, in the water, on land. Plastic pollution was everywhere; there was no getting away from it. There is no 'away.'"

After unloading on her waiter, she decided to be more strategic and start The Last Plastic Straw. "Basically what we are asking you to do is DO LESS . . . less consumption, less waste, less straws. It's a win-win!" she says. She made an invitation to bars and restaurants to be part of her movement to eliminate plastic pollution from the source. By simply stating "Straws available upon request" on menus, bars and restaurants can become part of the solution.

Thanks to Nunez, restaurants, bars, and cities from London to Miami are banning plastic straws voluntarily. When you return yours to your server, you should politely instruct them to

- provide a straw only when requested by a customer;
- provide either compostable or reusable straws; or
- get rid of straws completely.

On April 19, 2018, ahead of Earth Day, a proposal to phase out single-use plastics was announced during the meeting of the Commonwealth Heads of Government, a biennial summit of the heads of government from all Commonwealth nations. This includes plastic drinking straws and cups. It is estimated that as of 2018, about twenty-three million straws are used and discarded daily in the UK alone. Add to that India, Australia, Canada, and the other forty-nine members of the Commonwealth and you have a big source of plastic pollution, but one that they are now resolved to do something about. And the alternatives are literally grassroots.

A few months before that announcement, Queen Elizabeth II banned plastic straws and other one-use plastic items from her palaces.

Canada had already been planning on banning straws nationwide after 70 percent of voters polled endorsed a plastic straw ban.

Some thirteen thousand schools, workplaces, and event venues will be free of plastic bags and stirrers by 2019, thanks to a new push by food service company Sodexo, following similar steps by food service giants Aramark and Bon Appétit. In addition to bags and stirrers, Sodexo plans to phase out Styrofoam containers by 2025. The move, says a representative from the company, will eliminate 245 million single-use items that would have otherwise been used at its locations.

How hard would it be, after all, to go back to paper or to reeds such as hemp and papyrus? That is the subject of studies underway by straw makers as the pressure of single-use plastics bans forces their largest customers to scramble.

In 2018, *Business Insider South Africa* tested five alternatives to plastic straws: stainless steel, etched copper, glass, bamboo, and Khanyiso reed. All are reusable, and two are both biodegradable and renewable. Prices for each straw ranged between fifteen and ninety-five times its plastic counterpart. Metal straws, both copper and, to a lesser extent, stainless steel, had serious problems with heat because they heated or cooled to the temperature of the drink, which made them harder to handle or sip from. The bamboo straw left a bad taste, and the reed straw was nearly as bad. With the reed, everything hot tasted woody quickly, and anything cold tasted woody eventually. Bamboo left a foul green aftertaste and ruined the flavor of coffee. Glass had none of these problems and had the added advantage of being see-through. But glass could not be carried around safely. *Business Insider* concluded, "So this glass straw is a clear winner with one important caveat: it requires a sturdy carry case. Most likely something made of rigid plastic, rather than the hemp sleeve its makers provide."

We will look more into the emerging alternatives to plastics in later chapters, but this example of fiber straws illustrates an important part of the problem. Replacement with natural products is not always a viable solution. The experience may be less satisfying or less fit for purpose than provided by the plastic product. The question then becomes one of making choices less for utilitarian reasons than for ethical or ecological ones. Because the price was more than likely set with some long-dead economic theorist's thumb on the scale, or because a consumer who receives a take-out container for no extra charge feels no economical obligation to preserve it, the choice to go toward biodegradable bioplastic will most likely be made for reasons of conscience.

SOME QUICK FACTS ABOUT PLASTIC POLLUTION

Annual polymer production has grown from fifteen million tons in the 1960s to over four hundred million tons now and is expected to triple that number by 2050.

The quantity of plastic in the ocean is expected to nearly double to 250 million metric tons by 2025.

Ninety-nine percent of all plastic is produced from fossil fuels. While greenhouse gas emissions from plastics themselves are not significant, production of fossil fuels—especially from drilling, refining, and transporting "unconventional" discoveries as the "conventional" sources are depleted—has a very large impact on climate.

Estimates of how much global fossil fuel use goes to plastics range from 4–10 percent per year.

The ocean will contain one ton of plastic for every three tons of fish by 2025, and by 2050 more plastic than fish, by weight.

Economic damage to commercial fishing caused by plastic amounts to at least $13 billion every year, roughly one-third of the economic damage caused to all sectors.

Five countries—China, the Philippines, Indonesia, Vietnam, Thailand—together account for 55–60 percent of the total plastic waste going into the environment. The cost of ocean plastics to the tourism, fishing, and shipping industries is $1.3 billion in the Asia-Pacific region alone.

The amount of municipal waste produced on average by each European citizen is projected to increase from 520 kilograms in 2004 to 680 kilograms by 2020, an increase of 25 percent.

Today, 95 percent of plastic packaging material value—$80 billion to $120 billion annually—is lost to the economy after a short first use.

The recycling rate for plastics, in general, is even lower than for plastic packaging, and both are far below the global recycling rates for paper (58 percent) and iron and steel (70–90 percent).

Health Surprises

Recent technological advances permit new formulations at microscopic or atomic scales. The plastics industry is a leader in nanotechnology innovation. It is estimated that by 2020 the share of nanocomposites among plastics in the United States will be 7 percent, including materials that are reinforced with nanofillers (nanoclay and nanosilica) for weight reduction, nanocarbon for improved mechanical strength, and nanosilver as an antimicrobial agent in food packaging and medical products.

Like the consequences of the large-scale plastics experiment begun in the previous century, the consequences of nanoplastics are similarly left to be discovered and to being externalized. A team of Japanese researchers found some 40 percent of ocean fish caught for food had microplastics in their digestive systems. It is estimated from computer models that the oceans now contain some fifty-one trillion microplastic particles and another ten million tons of macroplastics enter the sea annually, to be slowly broken into smaller and smaller pieces by the effects of salt, sunlight, and agitation.

In 2004, Richard Thompson of the University of Plymouth, UK, analyzed the micro-debris on the beaches and waters in Europe, the Americas, Australia, Africa, and Antarctica. Thompson and his associates found that plastic pellets from both domestic and industrial sources were being broken down into pieces having a diameter smaller than a human hair. Thompson said there might be three hundred thousand plastic items per square kilometer of sea surface and one hundred thousand plastic particles per square kilometer of the seabed.

Another study collected samples of polyethylene pellets from thirty beaches in seventeen countries and analyzed them for micro-pollutants. Pellets from beaches in America, Vietnam, and southern Africa contained compounds from pesticides that had migrated into the polymeric chain. Other pellets contained cancer-causing and ecosystem-disruptive DDT and PCBs.

Ten years ago, Holger Koch of the German government's occupational safety office and Antonia Calafat of the US National Institutes of Health began looking at hospital monitoring data that was then becoming available from Germany and the United States. They saw a chemical called bisphenol A (BPA) and phthalates, two common ingredients of many polymers, appearing in urine samples for all parts of those two populations. They concluded that daily phthalate intakes might be substantially higher than previously assumed and even close to or exceeding thresholds previously observed for toxic effects in laboratory animals. "The toxicological significance [for] susceptible subpopulations (e.g., children, pregnant women) . . . remains unclear and warrants further investigation," they wrote in 2009.

Both Koch and Calafat continued to pursue those investigations. With her name appearing on more than twenty articles in the peer-reviewed literature in 2018 alone, Antonia Calafat has done everything she can to raise the alarm. Her studies now show that microplastics in the blood of pregnant women cross the placental barrier and directly result in embryonic developmental disorders, gestational diabetes, decreased birth weight, allergic asthma, and other respiratory problems in newborns. Worse, microplastics can be transmitted through mother's milk, meaning that infants who may already be adversely impacted receive an even higher dose at a most critical period in their development.

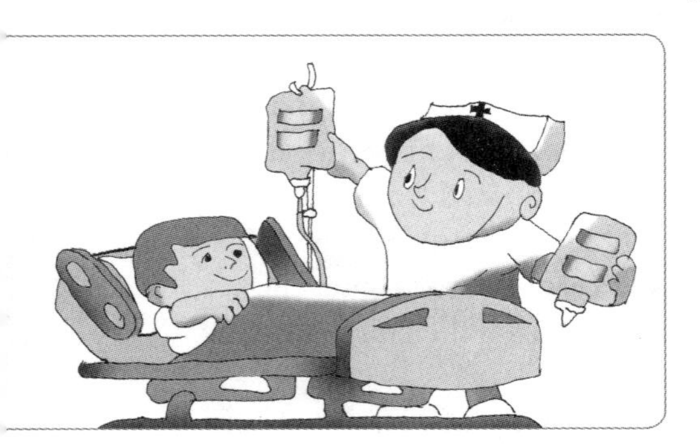

A century of plastic design improvements now let us keep our foods fresher for longer periods, provide us timed-release pharmaceuticals and non-degrading biomedical implants, and can prevent electronics and other household items from starting or spreading fires. But for each of these benefits, there are counterweighing human health risks related to exposure. We now know that some of the same chemicals used in plastics to provide beneficial qualities also act as endocrine-disrupting compounds (EDCs) that lead to problems in human and other populations.

In men, environmental or occupational exposures to EDCs can lead to declined reproductive capacity or possibly increased risk of testicular or prostate cancer. In women, exposure may give an increased risk for endometriosis, reproductive and other endocrine-related cancers, or impaired oocyte competence, ovarian function, or menstrual cycling. Effects of early life exposures may lead to altered sex differentiation, effects on neurological and reproductive development, and increased risk of reproductive problems or cancer later in life. Testicular dysgenesis syndrome can afflict males in utero or in infancy, later showing up as disturbed gonadal development, including cryptorchidism, hypospadias, and smaller reproductive organs, as a reduction in semen quality and infertility, and as an increased risk for testicular cancer.

PHTHALATES

The diesters of 1,2-benzenedicarboxylic acid (phthalic acid), commonly known as phthalates, are a group of man-made chemicals widely used in industrial applications. High-molecular-weight phthalates are used as plasticizers in flexible vinyl, which in turn is used in consumer products such as credit cards, flooring and wall coverings, food containers, medical implants, and window frames. Low-molecular-weight phthalates are in personal-care products (cosmetics, lotions, and perfumes) and in coatings, lacquers, solvents, and varnishes. They are also used to provide timed releases in some oral and subdermal pharmaceuticals.

As a result of all these consumer products, human exposure to phthalates is widespread. Skin contact is enough. For those identifying as men, it might come from cologne or aftershave. For those identifying as women, it might be from skin lotion or lipstick. For infants and children, mouthing fingers after handling plastic toys or food packaging can lead to higher phthalate exposures. So can breast milk, infant formula, and cow's milk, according to studies. Opting for almond, coconut, or rice milk won't save your child if that cardboard carton has a plastic liner or cap.

In newborns, the amount of phthalates in umbilical cord blood directly correlates to a risk of premature birth. Among girls, phthalate concentration correlates with premature breast development and early-onset puberty. Other developmental effects: allergies, rhinitis, asthmatic reactions, and direct toxicity. In one case-control study from Sweden, phthalate concentrations in indoor dust for 198 children ages three to eight showed a strong association with allergic asthma and eczema in a dose-dependent manner. Another study in Bulgarian children produced similar results, where increased plastic in house dust proportionally related to wheezing and rhinitis. A study of preterm infants who were provided polyvinyl chloride (PVC) respiratory tubing showed higher rates of hyaline membrane disease, proportional to the phthalate exposure.

Significant associations have also been reported between urinary phthalate concentrations and increased insulin resistance and waist circumference. These findings provide preliminary evidence of a potential contributing role for phthalates in insulin resistance, obesity, and related clinical conditions.

BISPHENOL A

BPA is in the epoxy resins used to line food cans, older plastic baby bottles, some dental sealants and fillings, adhesives, protective coatings, flame retardants, water storage tanks, and supply pipes. It starts as part of a polymer, but with normal heat over time, it degrades into its small-chain monomeric form. In that form, BPA can leach from its source into adjacent materials, such as water (in the case of bottles, pipes, or tanks) or food products (such as from the lining of a box, can, or pouch). There is widespread BPA lingering in body fluids, bones, and organs of people. It can be found in over 90 percent of the US population, where 96 percent of pregnant women test positive for BPA in their urine. It is now in US women's follicular fluid, amniotic fluid, umbilical cord blood, and breast milk.

BPA's hormone-changing properties were known as early as 1936, and evidence for other biological activity, such as effects on thyroid function, soon followed. In one epidemiological study, serum BPA levels were reported to be associated with recurrent miscarriage. Investigators also reported higher rates of polycystic ovary syndrome. Multiple studies have associated BPA exposure with weight gain and linked it to cancer, diabetes, heart disease, genital malformations, insulin resistance, neurological disorders, thyroid dysfunction, and more. However, most studies to date have only addressed single chemicals or classes of chemicals, and there are limited data on the interactions between chemicals within a class or across classes. Chemicals may interact additively, multiplicatively, or antagonistically in what is commonly referred to as the "cocktail effect."

The health effects of ingested plastics are not just limited to phthalates and BPA. We know of ill effects from esters of aromatic mono-, di-, and tricarboxylic acids, aromatic diacids, and di-, tri-, or polyalcohols, and many other additives and composite materials. The exploration of these medical effects is still in its infancy, and few governments have shown any willingness to disturb the marketplace until it is more clear which does what to whom.

In the meantime, it is nearly impossible to take a prescription medication or even use an over-the-counter vitamin without encountering time-release coatings on capsules, plastic lids on plastic pill bottles, microbead plastic desiccant pouches, and even a (synthetic) cellophane wrap for tamper-proofing. You can tell the checkout clerk at the grocery store you won't need a plastic bag because you brought your reusable cloth bag, but you may find it difficult to avoid having skin contact with the plastic handle on the shopping cart or basket, the laminate on the checkout counter, the credit card in your wallet, or the shock-resistant cover on the mobile phone that you might use for digital payment. You will likely be unable to do anything to prevent yourself from inhaling the microplastics in the hairspray the clerk used that morning, absorbing some of the microplastic-contaminated tap water you use to rinse and prepare your fresh vegetables, or eating the microplastic particles absorbed into the food as it was grown.

As addictions go, this one is a real brute.

CHAPTER 3

Alphabet Soup

If it can't be reduced, reused, repaired, rebuilt, refurbished, refinished, resold, recycled, or composted, then it should be restricted, redesigned, or removed from production.

PETE SEEGER

You can think of a polymer as a long backbone of repeating "vertebrae" with "rib bones" of monomers or shorter polymers that provide distinctive characteristics. We are going to run along that backbone now and look at the chains of molecules that might be dangling off to one side or another. To customize the properties of a plastic, different molecular groups "hang" from this backbone. It is the variable structure of these side chains that gives plastics so many possible properties. Acrylics, polyesters, polyurethanes, PVC, and silicones have similar backbones but different appendages. The silicone that gives sticky gloves a coefficient of friction 20 percent better than a bare hand in order to let football players pull in spectacular one-handed grabs is only a slight variation from the silicone grease depended upon as a friction-reducing lubricant for bicycles and automobiles. It's all in the appendages.

Other classifications of plastics are based on the chemical process used in their synthesis, such as condensation, cross-linking, and polyaddition, or the qualities that are relevant for manufacturing or product design, such as thermoplastic and thermoset, elastomer, structural, biodegradable, or electrically conductive. Plastics can also be ranked by various physical properties, such as density, tensile strength, glass transition temperature, and resistance to various chemicals and environments.

There are three major divisions: thermoplastics ("plastic" as in malleable) that soften and melt if enough heat is applied, such as polyethylene, polystyrene, and polytetrafluoroethylene (PTFE); elastomers (as in "elastic") that are heat-resistant but flexible; and thermosets ("set" as in final) that do not soften or melt no matter how much heat is

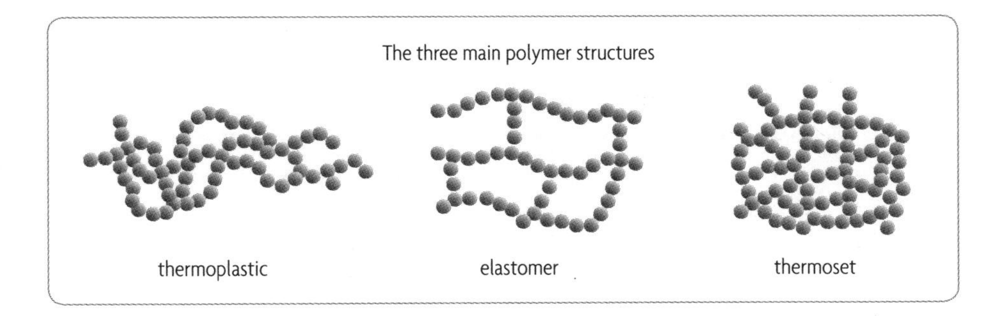

applied, such as the epoxy resins used in caulks; phenolic resins found in tool handles, billiard balls, and sprockets; and polyester resins such as reinforced fiberglass.

Most plastics contain *organic* polymers, which is to say they include *carbon*. This carbon is essential in the context of solutions to climate change and to decomposition of plastic, so we will circle back to this later.

Both the backbone and the side chains comprise many repeat units, formed from monomers. Each polymer chain will have several thousand repeating units. The vast majority of plastics are composed

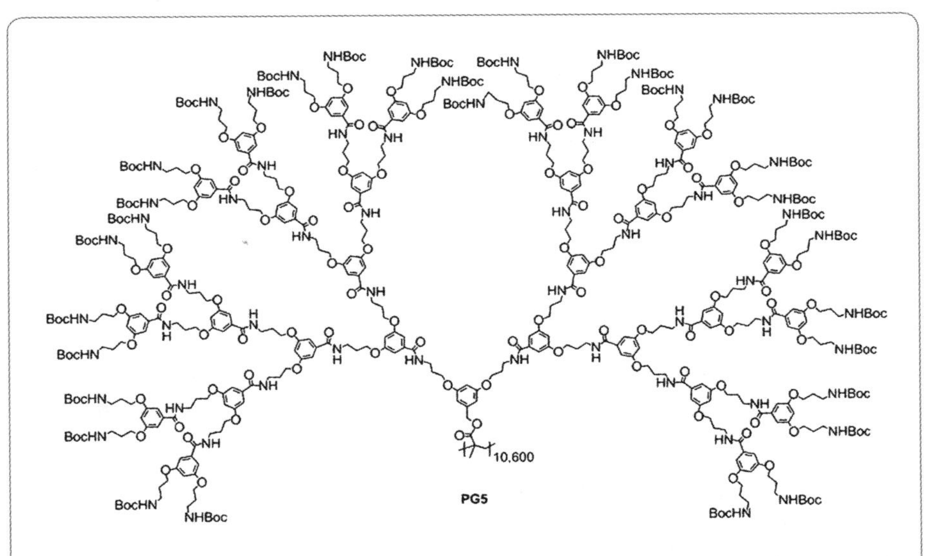

PG5, the largest stable synthetic polymer ever made, with a diameter of ten nanometers and a mass equal to two hundred million hydrogen atoms, required 170,000 bond formations.

of polymers of carbon and hydrogen alone, although oxygen, nitrogen, chlorine, silicon, and sulfur are also common.

Some plastics are partially crystalline and partially amorphous in molecular structure, giving them both a melting point (the temperature at which the attractive intermolecular forces are overcome) and one or more glass transitions (temperatures above which the extent of localized molecular flexibility is substantially increased). This determines relative flexibility or brittleness. Polyethylene, polypropylene, polyvinyl chloride (PVC), polyamides (nylons), polyesters, and polyurethanes are semicrystalline, semiflexible plastics. Others are entirely amorphous, and relatively inflexible, such as polystyrene and its copolymers, polymethyl methacrylate, and all the class of thermosets.

You have probably seen these letter symbols:

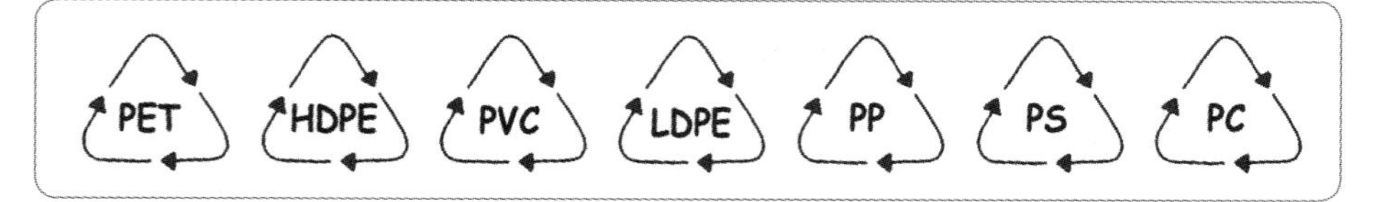

Or maybe you have seen a numeric version:

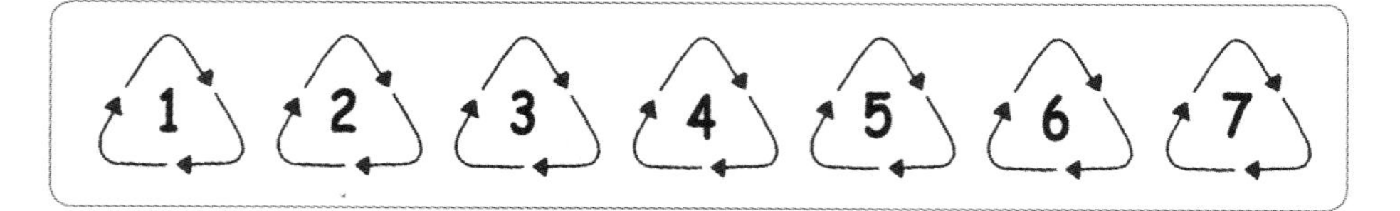

PET, PETE, PETG, and PET-P (polyethylene terephthalate), the most common types of plastics, first appeared in the postwar era with names such as polyester or Dacron. Today these plastics are common in peanut butter jars, water bottles, and containers for salad dressings, cooking oils, mouthwashes, and window cleaners. Since carbonation tends to attack plastics, PETs have become a popular material for holding Coca-Cola and other carbonated drinks or for acidic beverages, such as fruit or vegetable juices. The owner of the Dacron trademark, INVISTA, makes a range of PET pillows and mattress fibers, including Memorelle, Silky Soft Fiber, Supreme Loft, CoolFX Fiber, and CleanFX Fiber. PET is also used for making mechanical parts, food trays, tennis balls, and other items that have to endure abuse. Because it can be transparent, PET films are popular in produce wrappers, stand-up pouches, and microwavable packaging.

PE (polyethylene), sometimes known as polythene or polyolefin, was discovered in 1933 and has evolved into multiple forms, including low-density polyethylene (**LDPE**) and high-density polyethylene (**HDPE**). Polyester (**PES**) is in fibers and textiles. PEs are cheap, flexible, durable, and chemically resistant. LDPE is used to make films and packaging materials, garbage bins and bags, outdoor furniture, siding, floor tiles, shower curtains, and clamshell packaging. HDPE is used for shampoo containers or milk bottles, plumbing, and automotive fittings. While PE has low resistance to chemical attack, a PE container can be made much more robust by exposing it to fluorine gas, which modifies the surface layer of the container into a much tougher polyfluoroether (**PFE**), such as polytetrafluoroethylene (**PTFE**).

PVC (plasticized polyvinyl chloride) is what is used to make credit cards, plastic pipe, plumbing pipes and guttering, electrical wire/cable insulation, shower curtains, vinyl siding, window frames, flooring, and cordial, juice, or squeeze bottles. Unless you are part of an undiscovered tribe in the Amazon or are living in Papua New Guinea, chances are you contact PVC on a daily basis. A thin film variant, **PVDC** (polyvinylidene chloride), was used for food packaging, such as Saran Wrap, but has since been replaced with polyethylene by Saran.

PP (polypropylene), standard in lunch boxes, takeout food containers, and ice-cream containers, was discovered in the early 1950s and separately invented about nine times. The patent litigation was not resolved until 1989. Polypropylene is similar to its ancestor, polyethylene, and shares polyethylene's low cost, but it is much more robust. It is used in everything from bottle caps, drinking straws, and yogurt containers to appliances, car fenders, and plastic pressure pipe systems.

PS (polystyrene) is found in foam peanuts; food containers; plastic tableware; disposable cups, plates, and cutlery; and compact-disc and cassette boxes. Variations include acrylonitrile-butadiene-styrene (**ABS**) found in electronic equipment (computer monitors, printers, and keyboards) and drainage pipe; polycarbonate/acrylonitrile-butadiene-styrene (**PC/ABS**), a blend of PC and ABS that creates a stronger plastic used

TABLE 1. Melting points of plastics

POLYMER TYPE	MELTING POINT*	
PVC	210–500°F	99–260°C
LDPE	230–248°F	110–120°C
HDPE	250–350°F	121–177°C
PC	302–500°F	150–260°C
PP	320–338°F	160–170°C
PS	347–464°F	175–240°C
PET	500°F	260°C

*Melting points vary according to the density and exact formulation of the polymer.

in car interior and exterior parts and mobile phone bodies; and polyethylene/acrylonitrile-butadiene-styrene (**PE/ABS**), a slippery blend of PE and ABS used in low-duty dry bearings. High impact polystyrene (**HIPS**) is in refrigerator liners, food packaging, and vending cups.

PU (polyurethane) was invented in 1937 and can be found in the blown foam for mattresses, furniture padding, and thermal insulation. It is also one of the components of spandex and currently the sixth or seventh most commonly used plastic.

PA (polyamides or nylons) is in fibers, toothbrush bristles, tubing, fishing line, and low-strength machine parts, such as engine parts or gun frames.

PC (polycarbonate) is in compact discs, eyeglasses, riot shields, security windows, Formula One race cars, traffic lights, and bulletproof glass.

PCE (perchloroethylene or tetrachloroethylene) is the most popular solvent in conventional dry cleaning. The chemical is a synthetic volatile organic compound (VOC). If a compound is volatile, it can easily and quickly vaporize. PCE vaporizes at room temperature, and this is how dry cleaning works. However, once the solvent evaporates, it is easily inhaled. Because of this, both dry cleaning employees and customers are at risk of inhaling a toxic chemical.

PEEK (polyetheretherketone) is part of the polyaryletherketones (**PAEK**) family and has the best balance of heat resistance, wear resistance, and chemical resistance among all thermoplastics. Its biocompatibility allows for use in medical implants and agricultural applications. One of the most expensive commercial polymers, it can be found in the aerospace moldings used on the Space Shuttle.

PEI (polyetherimide) is produced by the General Electric Company under the Ultem trademark. It is transparent and strong and has inherent flame resistance. It can be injection molded at temperatures of 150–350 degrees F, while similar plastics require 660–800 degrees F. At 356 degrees F, its tensile strength and flexural modulus are 6,000 and 300,000 psi, respectively. It can be made into any shape or into transparent sheets as thin as 0.25 mil, making it useful for eyeglasses and shatterproof glass.

PF (phenolics or phenol formaldehyde) is a high-modulus, relatively heat-resistant, and exceptionally fire-resistant polymer used for insulating parts in electrical fixtures, paper-laminated products (Formica), and thermally insulating foams. This is the original Bakelite that can be cast into any shape but only in colors of red, green, or brown. It is a thermosetting plastic, impossible to recycle.

PLA (polylactic acid) is a biodegradable thermoplastic derived from lactic acid and can be made by the fermentation of various agricultural products, such as cornstarch and dairy products. It is used in a wide variety of products, from paints and printer filaments to packaging and pallets.

PMMA (polymethyl methacrylate, acrylic polyepoxide, or "epoxy") is used as an adhesive, in contact lenses (the original "hard" variety), glazing, and rear light covers for vehicles. It forms the base of decorative and commercial acrylic paints. Composites might include glass-reinforced plastic (fiberglass, Acrylite, Perspex, Plexiglas, and Lucite), in which the structural element is glass fiber, and carbon-epoxy composites, in which the structural part is carbon fiber (Teslas and aircraft cockpits). Makrolon is a high-impact polycarbonate plastic made by Bayer. Lexan is a high-impact polycarbonate plastic made by Saudi Basic Industries Corporation (SABIC).

PSU (polysulfone) is a high-temperature, melt-processable resin used in membranes, filtration media, water heater dip tubes, and other high-temperature applications. Polyphenylsulfone (**PPSU** or Radel) is the highest-performing sulfone polymer, with better impact and chemical resistance than PSU and PEI. Radel's R-7000 Series products are specially formulated for use in aircraft interiors to meet Federal Aviation Administration regulations for heat release, smoke, or toxic gas emissions.

PTFE (polytetrafluoroethylene or Teflon) is found in heat-resistant, low-friction coatings, such as nonstick surfaces for frying pans, plumber's tape, and water slides. It was discovered during World War II as a secret part of the Manhattan Project's process to refine uranium for the first atomic bomb.

A CAUTIONARY NOTE ABOUT TEFLON

Nonstick pans are a wonderful invention, saving cooking oil, preserving essential flavors in ingredients, and easing the chores of dishwashers. It is small wonder that nonstick pans are now 90 percent of the aluminum pan market. But we need to be cautious. The reason that foods don't stick to nonstick is fluoropolymers, but at 500 degrees F (260 degrees C) those can split into smaller plastic fragments. The coating begins to break down, and toxic particles and gases, some of them carcinogenic, are released. It is sometimes possible to see when this has happened because the surface degrades, losing its characteristic smoothness and making it more difficult to clean. But even before that happens, the polymers can break down at the molecular level, and you wouldn't necessarily see any damage.

At very high temperatures—660 degrees F and above—coatings give off invisible fumes strong enough to cause polymer-fume fever, a flu-like condition indicated by chills, headache, and fever. At 680 degrees F, Teflon releases at least six toxic gases, including two carcinogens, according to a study by the Environmental Working Group, a nonprofit watchdog organization.

The popular magazine *Good Housekeeping,* after commissioning its own study, made these recommendations:

> Any food that cooks quickly on low or medium heat and coats most of the pan's surface (which brings down the pan's temperature) is unlikely to cause problems; that includes foods such as scrambled eggs, pancakes, or warmed-up leftovers. And many other kinds of cooking are safe as well: In [our] tests, the only food prep that yielded a nonstick pan temperature exceeding 600 degrees F in less than ten minutes was steak in a lightweight pan. But to be cautious, keep these tips in mind:
>
> **Never preheat an empty pan.** Each of the three empty nonstick pans we heated on high reached temperatures above 500 degrees F in less than five minutes—and the cheapest, most lightweight pan got there in under two minutes. Even pans with oil in them can be problematic; our cheapest pan zoomed to more than 500 degrees F in two and a half minutes.
>
> **Don't cook on high heat.** Most nonstick manufacturers, including DuPont, now advise consumers not to go above medium. To play it safe, set your knob to medium or low and don't place your nonstick cookware over so-called power burners (anything above 12,000 BTUs on a gas stove or 2,400

watts on an electric range); those burners, seen more often in recent years, are intended for tasks such as boiling a large pot of water quickly.

Ventilate your kitchen. When cooking, turn on the exhaust fan to help clear away any fumes.

Don't broil or sear meats. Those techniques require temperatures above what nonstick can usually handle.

Choose a heavier nonstick pan. Lightweight pans generally heat up fastest, so invest in heavier-weight cookware—it's worth the extra money.

Avoid chipping or damaging the pan. To prevent scratching, use wooden spoons to stir food, avoid steel wool, and don't stack these pans. (If you do, put a paper towel liner between them.) How long can you expect your nonstick cookware to last? DuPont's estimate, based on moderate usage, is three to five years. Some experts advise replacing your nonstick cookware every couple of years. What should you do if the pan does become damaged? A clear answer: Throw it out.

Another danger of nonstick pans comes from a chemical used to manufacture the fluoropolymers, **PFOA** (perfluorooctanoic acid), which is associated with tumors and developmental problems. After paying $343 million to settle a lawsuit in 2004, DuPont agreed to phase out PFOA, and did by 2015, but many older pans may still have the coating. PFOA is still on the market, however, in microwave-popcorn bags, fast-food packaging, shampoo, carpeting, and clothing. Studies show that most of us have PFOA in our bloodstreams, and babies show trace amounts at birth.

MF (melamine formaldehyde) is one of the aminoplasts, used as a multi-colorable alternative to phenolics, for instance in moldings (e.g., break-resistant alternatives to ceramic cups, plates, and bowls for children) and the decorated top-surface layer of the paper laminates (Formica).

UF (urea-formaldehyde) is another of the aminoplasts, used as a multi-colorable alternative to phenolics as a wood adhesive (for plywood, chipboard, hardboard) and in electrical-switch housings.

Furan is a resin based on furfuryl alcohol used in foundry sands and biologically derived composites.

Polyimide is a high-temperature plastic used in materials such as Kapton tape.

Plastarch material is a biodegradable and heat-resistant thermoplastic made from cornstarch.

Silicone is a heat-resistant resin used mainly as a sealant but also for high-temperature cooking utensils and as a base resin for industrial paints and epoxies.

MX is the recycling symbol for a mixture of any of the above.

Breaking Bad Bonds

As we have seen from this brief overview, to call something "plastic" is a very nonspecific term. The distinctions between one kind and another matter even more when we are trying to solve the problems of reuse, recycling, or biodegradation. Perhaps it would be better if we used emojis instead of letters or numbers.

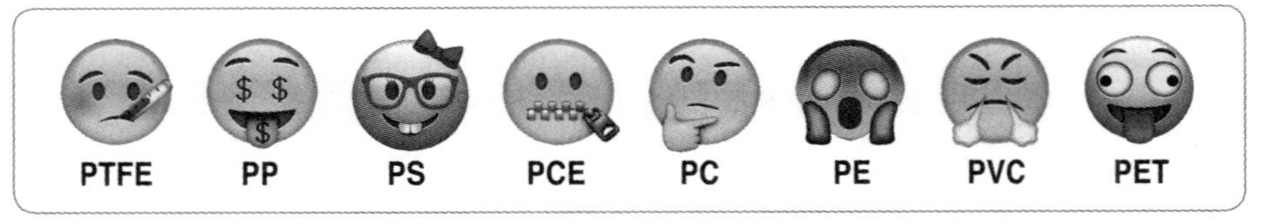

Let's look at some of the more common varieties and get under the hood to try to see how they are made and potentially unmade. Thanks to Michael Tolinski's scholarly work *Plastics and Sustainability: Towards a Peaceful Coexistence between Bio-based and Fossil Fuel-based Plastics*, this very difficult subject has become much more accessible to lay readers.

THERMOPLASTICS

Because they can be melted, usually at relatively low temperatures, thermoplastics offer more avenues to destruction than elastomers or thermosets.

PE is most often made by oil and gas refineries that crack ethane to make ethylene using steam. Catalysts are used to dangle pendant hydrogen from the carbon backbone. In Brazil, this process has now been adapted to sugarcane, although at a higher cost for the moment. Raw sugarcane is pressed to syrup, microbes digest the sugars to produce ethanol and CO_2, and the ethanol is then dehydrated and polymerized to make PE. Other sugary plant stocks are being investigated, as are new microbes that might be capable of breaking sugars down.

Can microbes break down PE? So far they haven't been all that inclined, but researchers are looking into additives that might make HDPE and LDPE more appetizing to them. Since these categories include blown-film plastic shopping bags and many single-use containers, biodegradable PE would be a major breakthrough.

In the meantime, PE can be easily melted if the time and energy are spent to collect, transport, clean, segregate, and liquify it. Unfortunately, the by-products of melting PE are CO_2 and water, both major greenhouse gases.

PP is a polymer of propylene assembled from naphtha, a by-product of oil refineries, in much the same way as PE is made. The pendants are methyl chains (CH_3) attached to alternating carbon atoms in the backbone. When the methyl groups all line up on the same side of the backbone, it is called an isotactic polypropylene. If they alternate, it is called syndiotactic, and if they line up at random, it is called atactic. Each arrangement has different properties, but unless filled, blended, or pigmented, PPs are easily recycled.

PVC is similar to PP except instead of methyl groups, it attaches chlorine atoms. This means that factories producing PVC can be dangerous workplaces. Hydrogen chloride, a toxic gas, is produced when ethylene and chlorine are combined. When ethylene dichloride is cracked to produce vinyl chloride, it can become hydrochloric acid on contact with water. Since water is in body tissue, vinyl chloride exposure can cause pulmonary edema, circulatory system failure, and death. Chlorinated dioxins, also produced by this process, are slower-acting poisons. High doses of toxic plasticizers (up to 30 percent by weight) are used to soften the PVC, and those can and do leach, both before and after the PVC gets to market.

For this reason, flexible PVCs are not considered recyclable. Unplasticized PVCs can be recycled, but recycled PVC costs much more than virgin PVC, so no one recycles it. Incineration releases deadly hydrogen chloride, so that's not an option. That's also why choosing PVC for building materials is a really bad idea, especially if you happen to be a firefighter. Unless they can be reused, PVCs are landfilled for safety reasons.

PS has a backbone with a repeating pattern of two carbon atoms, one of which has a side pendant with a constituent other than hydrogen. The production process is nasty, requiring the production of carcinogenic benzene and styrene. PS is brittle but can be grafted to rubber to make a high-impact PS or HIPS, or it can be blown into foams to make expanded polystyrene (EPS), which is a very efficient insulator and heat absorber. Neither PS nor EPS can be biodegraded or recycled, which makes their use for items such as vending machine cups and clamshell containers particularly offensive.

Kathleen Draper, my coauthor of *BURN: Using Fire to Cool the Earth,* discovered by experimentation that EPS dissolves quickly in acetone at room temperature, and the lightweight, gummy product can then be blended with up to 60 percent biochar and molded into virtually any shape, which hardens in a few days. It could be recast for roofing tile, surfboards, or outdoor furniture, thereby sequestering photosynthetic carbon away from the atmosphere and oceans indefinitely.

PET is a condensation polymer, meaning that two different materials react to form the long backbone with alternating monomer segments. In the case of PET, the reactants are ethylene glycol and terephthalic acid or dimethyl terephthalate (DMT). These components can come from either fossil fuels or bioethanol.

After Coca-Cola introduced the PET PlantBottle, made from 30 percent sugarcane, PepsiCo announced it will start using beverage bottles that are 100 percent bio-based PET, made from food wastes such as potato and orange peelings. That is a healthy competition. The energy requirements and emissions for producing bio-PET are nearly half of that required for making fossil PET.

While bio-based doesn't mean biodegradable, PET and HDPE are the most recycled of plastics in the world today, and some 22 percent goes back to food and beverage containers while the remainder is downcycled into carpets or other products. The challenges for recycling are the water required (the flakes have to be thoroughly washed) and the difficulty of getting people to do it.

Nylon polyamides are easy to mold and make into fibers, having a nitrogen-based amide group sharing the backbone with atoms of carbon. While they began life as a petrochemical polymer, increasingly nylons are being converted to plant-based sources, notably castor oil, with 50 percent less greenhouse gas emission in manufacturing. Post-consumer nylon requires very careful separation to be recycled because it is widely used as a cover or fill and may have become contaminated with less-recyclable products.

PC (polycarbonate) is commonly used in optical storage media, eyeglasses, shatterproof windows, and airliners. It is light, strong, and temperature tolerant. The process of synthesizing it is dangerous, however, because it is produced by the reaction of phosgene (a chemical weapon) and bisphenol A (an endocrine disruptor). Like PVC, landfilling could be the safest end-of-life resting place.

THERMOSETS

Thermosets, such as epoxies, polyurethanes, unsaturated polyesters, and styrene resins, can't be remelted, so they don't go well with most recycling systems, although they can sometimes be chemically reacted to dissolve. Scrap thermoset may be ground up and used in highways, buildings, or as product fillers, but since it doesn't biodegrade, setting it loose in the environment is not ecologically wise. Like PC and PVC, these substances are best collected and landfilled.

BIOPLASTICS

In the world of plastics, the emerging biopolymer sector is the place to be. Annual demand growth of about 11 percent over the past decade could give way to even faster growth in the coming years, driven by concern about climate change and other negatives of the fossil economy. Starch-based plastics are leading the switch because they are biodegradable as well as plant-based, but industry experts say bio-based plastics of all types could theoretically replace 90 percent of traditional polymers.

That should not be hard to understand because up until about a hundred years ago all human-use polymers were of natural origin, made from bone, skin, shells, and plant fibers. Celluloid and cellophane are natural materials. Casein resins and paints are based on milk protein.

Starch polymers are usually derived from corn, wheat, and potatoes, which are first gelatinized with heat and moisture and then extruded. Glycerols, amides, and citric acid are added to control the starches' melting temperature. When Henry Ford made an entire Model T from soy polymer, he demonstrated its strength by slamming an ax into the body. The ax bounced off, doing no harm to the car. Soy proteins have the tendency to absorb water, so it is probably a good thing that Ford was not building boats. That is the takeaway from working with nature—each feedstock has its own character, and you need to learn to use the one most fit for purpose.

PLA, another polymer resin from milk, will not degrade in most outdoor environments but will readily break down in composting conditions above 60 degrees C. PLA is replacing metallic polymers used for snack bags, such as Frito-Lay's SunChips. Tests have shown the SunChips bags compost in 180 days and carry no residual toxicity to soils. However, because its biodegradable quality can gum up recycling lines and because it cannot be mechanically separated from PET bags that look and weigh the same (near-infrared spectroscopy is required), PLA has been banned from recycling in some jurisdictions until these issues are resolved.

Another class of biopolyesters related to PLA is **PHB** and related copolymers. These are created by fermentation processes of bacteria and enzymes that digest glucose and hold PHB as granular inclusions within their cell bodies, with these stored plastic precursors becoming as much as 29 percent of the organisms' weight. PHBs are then harvested from the bacterial cells and refined by epoxidation, block copolymerization, and annealing. This unusual process has shown potential for use in medical implants, tissue grafts, and drug delivery. "Bacterioplastics" have been converted for use in shopping bags, shampoo bottles, and carpets, but the cost of manufacturing makes them less competitive than existing products for the time being.

One attribute of PHB is particularly endearing. Tests of PHBV film show that it dissolves in natural seawater and gets decomposed by benthic microbial activity, with 75 percent biodegradation in thirty-five days. A throwaway plastic that naturally decomposes in seawater is the Holy Grail.

Likely one of the creatures helping to break down the PHBV in that trial was marine microalgae, itself a potential source for future plastics. Algae is a solar-powered factory for producing renewable biomaterials. It grows fast, doubling in biomass daily when food and growing conditions are right. It requires less land, less energy, and fewer inputs than corn or soybeans and, depending on additives, can be completely biodegradable.

The bottom line for all these types of plastics, their production, use, and end-of-life possibilities is that no single answer can fit all cases. Some plastics now in use are difficult to justify continuing to produce once their adverse health and environmental costs are fully and fairly tabulated. Some of those have ready substitutes that are more benign, some of which even come in at a lower price. When the price of the replacement is higher, we should ask whether all the costs of the original have been taken into account or whether we have externalized the health and ecosystemic damage just to make these products *appear* cheaper. When looked at in this way, many bioplastics that are presently unaffordable, have low recyclability, are difficult to compost, or have other problems might still have a lower long-term social, health, and environmental cost than products now in use.

Our biggest obstacle to finding solutions could be the confusion that has been designed into the entire industry. Calling something plastic does not tell you whether it is PLA, PET, or PVC. Even using two-, three-, and four-letter acronyms for long, molecularly descriptive chemical names only confuses most people, who likely don't know the difference between PLA and PET. If PLA is plant-sourced, does that mean it can go into the recycle bin with PET? (It can't.) The packaging industry continues to revise labeling standards, but so far these have only been minor tweaks on recognizable symbols that have a severe cognitive deficit from birth. Color-coding all plastics has been deemed too onerous for the industry, but very little short of that will make up for the limitations of human nature.

Barring some sort of massive collapse of industrial civilization, it seems unlikely that we will stop making and using PE, PVC, PC, or most of the thermosets anytime soon. For the plastics we can't refuse, reuse, recycle, keep using, break down, or compost, our best bet will be to collect and inter them in some well-designed sanitary landfill, there to remain for eternity or until some better solution appears.

CHAPTER 4

Recycling

Whenever you find yourself reaching for that single-use, villainous item, dare to ask yourself, "What will my plastic legacy be?" The plastic legacy concept isn't my shameful guilt-inducing plea to convince you not to do something that is commonplace in every single shop around the world; it's a weapon that you can apply to crush any fleeting, fuzzy feeling of apathy to the magnitude of our struggle with this addiction to plastics. It is a term for realizing the impact of each tiny, stupid package that you buy.

GEORDIE WARDMAN
OUR PLASTIC LEGACY: HOW TO QUIT PLASTIC, WANT LESS, AND LIVE GREEN DAILY

All polymers begin as sunlight that falls upon the earth. Billions of years ago, some single-celled organisms captured sunlight from green chloroplast cells before dying and settling to the bottom of the ancient ocean. Later, larger multi-celled algae and zooplankton did the same, for hundreds of millions of years. Under the heat and pressure of layers of sediment and shifting continents, their carcasses became a solid substance called kerogen and, eventually, the hydrocarbon chains we've come to know as natural gas, petroleum, and coal.

The first modern, widely available plastic was made from nitrocellulose, a plant material, and camphor resin, a coal tar. An American inventor named John Wesley Hyatt won a prize offered by a billiard manufacturer looking for a cheaper alternative to ivory. Hyatt's "celluloid" won with a bang, because occasionally, according to Hyatt, "the violent contact of the balls would produce a mild explosion like a percussion gun-cap." Nitrocellulose is explosive.

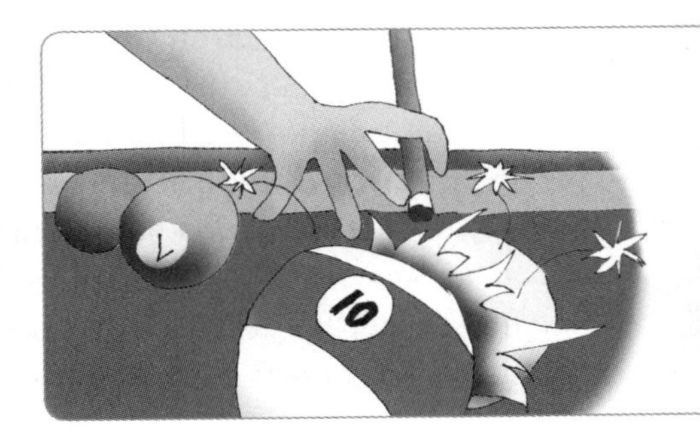

The birth of plastics in the early twentieth century made possible the birth of mass-scale injection molding—shooting liquid plastic into a closed mold, letting it harden (which can take only seconds), and then opening the mold to eject the finished product. As Heather Rogers tells us in "A Brief History of Plastic" for the *Brooklyn Rail*, "In the mid-1930s, at one company the same worker that formerly made 350 plastic hair combs per day could turn out more than 10,000 in equal time using injection molding."

Some long-chain hydrocarbons can and do get broken down by natural agents, such as sunlight, water, and bacteria, but those kinds of biodegradable molecules have not been the stock and trade of the plastic industry.

"The real trick is to make them stable when you're using them, and unstable when you don't want to use them," explained Marc Hillmyer, who leads the Center for Sustainable Polymers at the University of Minnesota. Instead of looking for durable, long-chain molecules, chemists are now looking for short-chain molecules that can be "stitched together" for use and then "unzippered" when ready to be discarded. XiaoZhi Lim, writing in 2018 for the *New York Times*, explains:

> In 2016, Dr. Hillmyer and his team made a polyurethane from unzipping polymers that were chemically recyclable. Molecular units derived from sugar linked up to make the polymers, which then cross-linked into polyurethane networks. The foam remains stable at room temperature but unzips into units if heated above 400 degrees Fahrenheit.

The economy of this approach leaves a lot to be desired because it means every polymer made this way has to go through the manufacturing process twice, first for assembly and second for disassembly. Product sales and marketing support the first stage, but what sustains the second? Even when the recovered short chains can be recombined, the cost of collection, transport, and recovery may be higher than stitching virgin polymers, and this doesn't factor in the need to retrain a studiously nurtured throwaway culture to get in the habit of returning its discards to their source.

Ever wonder why no nuclear plants, even those that melted down at Three Mile Island, Chernobyl, and Fukushima, have been decommissioned? It's because the process of disassembling and downcycling their component parts is more expensive than building the plant in the first place. As we shall see, the same can be true for plastics, or not, depending on many compounding factors.

Making Plastic Precious

In late 2018, as I searched for innovative approaches to the various plastics issues, I came across a twenty-three-year-old student, Dave Hakkens, who was making YouTube tutorials and a website, Making Plastic Precious, as part of his graduation project from a design school in the Netherlands. Hakkens was showing how to recycle HDPE plastic (milk bottles, bottle tops) into craft and engineering projects (plates, furniture, roofing shingles). In a string of well-made videos, Hakkens shows how, using only simple tools, anyone can make a decent living turning waste plastic into useful things.

Hakkens and his partners provided free plans for do-it-yourself machines that could be fabricated as easily in Kenya or Nepal as in Europe or North America. Every village in the world needs a shop like his. You can see videos and download his plans to make the tools at preciousplastic.com. His basic DIY setup includes the following:

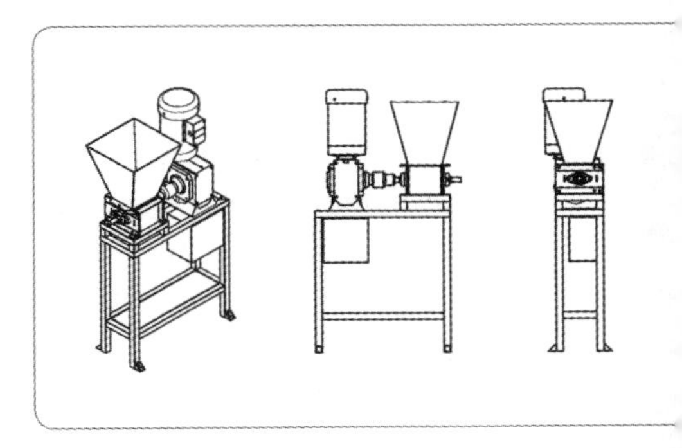

A shredder machine: Plastic waste is shredded into flakes to be used in other machines to create new things. Smaller flakes are easier to store and wash. The shredded plastic can be used as raw material for other machines or be sold back to the industry. Changing the sieve inside the shredder makes different patterns for different products. If you shred plastic by colors, you can have more control over the look and feel of your creations, adding value to the material.

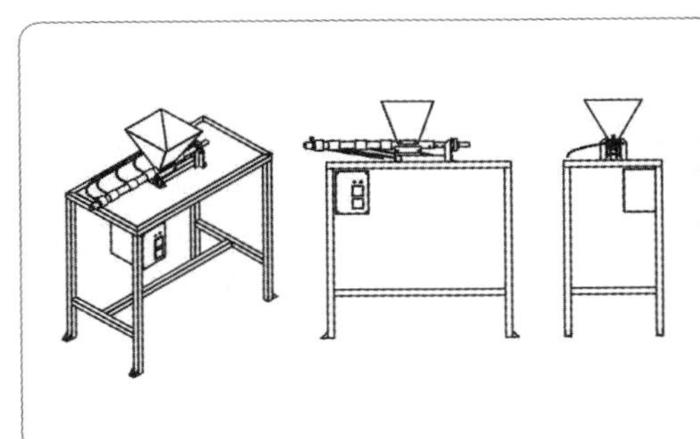

An extrusion machine: Extrusion is a continuous process in which plastic flakes move from a hopper into a flowing line that melts the flakes to mold into any shape as they harden,

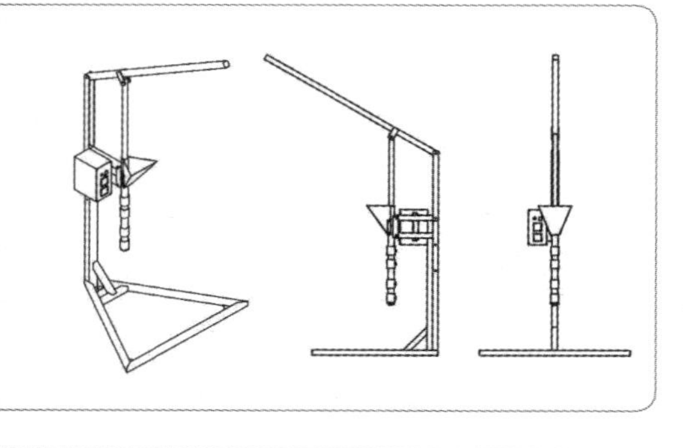

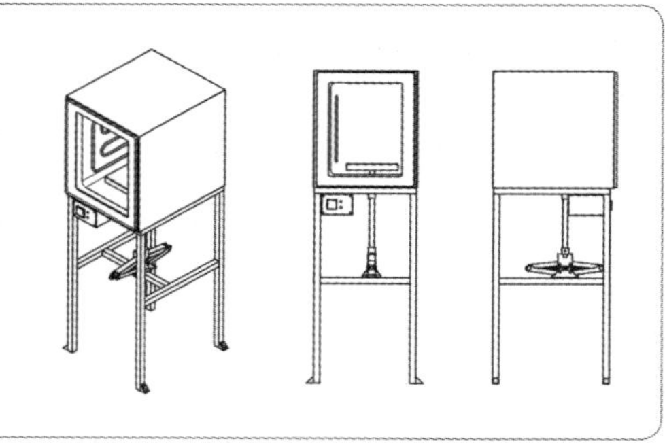

make 3D printing filament, or be layered in new and creative ways. This technique nicely blends differently colored plastics together and outputs a homogenous color or a smooth color gradient.

An injection machine: By injecting hot plastic flakes into a mold at high speed, you can make small objects in large volume.

A compression machine: Something similar to an electric kitchen oven heats the plastic, which flows into a mold where a car jack is used to apply pressure. The process allows for bigger objects, such as sheets of plastic, countertops, roofing tiles, and blocks, that can be further worked on to make more complex items, such as furniture and bicycle frames. With mixes of different-colored plastics, the results look like quarried stone.

The total energy for collecting, sorting, and reprocessing one thousand pounds of postconsumer PET and HDPE is less than one-sixth as much as that required for producing virgin resin (which inherently possesses great energy content in its chemical composition). The greenhouse gas emissions from recycling, including emissions from material collection, are about one-quarter to one-third of those in virgin resin production. However, solid waste production when recycling the postconsumer materials is higher than with virgin material production because of the residuals and unusable materials produced in the sorting and reprocessing steps. Hakken's kits have the advantage of working with clean, uniform materials. As feedstocks have to be reclaimed from mixed streams, the disadvantages increase.

One significant disadvantage is if biodegradable PLA or PHB get mixed into the feedstock stream. Since they look and weigh the same as PET and HDPE, it's an easy mistake to make, and they can get as far as the extruder before they begin to mess up the system. Even if they melt and homogenize into the resin, after the product cools, it may be weak or inferior in other ways, or it may spontaneously disintegrate.

This is the problem when biodegradable plastic bags, cups, and similar consumer items get mixed into the recycling stream. They gum up the works. They may be nearly indistinguishable from standard PET and polyolefins to sorting equipment, but their room-temperature degradation additives can compromise the quality of the entire stream.

This is ironic because biodegradability should be our goal in developing new plastics that are ecologically benign, but it conflicts with another goal of closing the resource-to-waste cycle. Worse, many so-called biodegradable polymers require carefully controlled composting, another space-, energy-, and labor-consuming task. Those who imagine these biodegradable plastics can just be rolled out in thousands of commercial products that replace non-bio plastics are assuming that large-scale industrial composting facilities actually exist. In most municipalities, they don't.

Another downside of bio-based polymers is that even if only 10 percent of all plastics were starch-based and derived from food crops, such as maize, soybeans, or sugarcane, it could drive up food prices, making them less affordable for the lowest-income segment of the population. Moreover, the life-cycle assessment doesn't give them as much of a greenhouse gas advantage as you might think. Biopolymers rely on fertilizers and pesticides for feedstock production, causing N_2O emissions, ozone depletion, acidification, toxicity, and eutrophication of water bodies as those agricultural chemicals leach from the soil. Chemical fertilizers give soils a short-term boost while killing them biologically over the long term, and this makes a nasty treadmill for farms that come to rely upon them. More chemicals are needed every year, while crops become weaker and sicker at the same time.

Pure traditional polymers, such as PP and PE, perform better on life-cycle analysis because of their relatively straightforward chemical processing from fossil fuels, but that is a different kind of treadmill and equally fragile in the long term. To put it simply, raw materials for producing plastics all have potential environmental impacts, whether through drilling for oil or fertilizing and harvesting crops for biofeedstock production, which currently still relies heavily on the use of fossil fuels.

For all the build-out of solar, wind, and other renewable energy sources over the past few years, the gain of energy being produced has been less than the growth in energy demand. A scientific study supported by the Dutch Ministry of Infrastructure warned in December 2018 that the renewable energy industry could be about to face a fundamental obstacle: shortages in the supply of rare metals. Rare earth metals are used in solar panels and wind turbines, as well as in electric cars and consumer electronics. We don't recycle them, and there's not enough to meet the demand.

The same is true for plastics that derive from either fossil fuels or agricultural crops diverted from your dinner table. The day of those feedstocks is ending. It must.

Barriers

What are the barriers to recycling plastic? Along with the cleanliness and overall quality of the incoming material, plastics have several particular qualities that interfere. Plastics began their lives as organic materials broken apart in extreme conditions in laboratories or factories. They can't just be melted or dissolved without altering their artificial molecular structure. In many cases, you have to disassemble the backbone and its dangling chain and rebuild from scratch.

Not everything is either possible or practical to recycle. Roughly half of the plastics in the municipal recycling stream are unsuitable.

Not all polymers are chemically compatible in a mixed recycling stream. Many theoretically recyclable plastics can't be reprocessed together because they differ in melting temperature, melt viscosity, or molecular structure. Thermoformed or injection-molded products are not PE film or PS foam. Foam has its own issues because its low density makes the costs of handling and storage hard for recyclers to justify.

The recycling stream may contain plastics that have degraded from use, heat, light, or outdoor weathering. Reprocessing damages molecular properties further, so that has to be compensated for by adding new stabilizers to restore some of the lost properties.

Even within some classes, such as PET, there can be differences that frustrate recycling. Even though they are both clear packaging materials, thermoformed PET products have to be separated from the bottle PET stream because of viscosity differences, label/adhesives, additives, and sorting difficulties (thermoformed PET clamshells can be difficult to distinguish from PS, PVC, and PLA clamshells).

It is often unclear what filler or fiber an old plastic part is loaded with. A common-looking PET bottle may be blow-molded from material with multiple ethylene-vinyl alcohol barrier layers, or it may be contaminated with milk, orange juice, paint, or kerosene. Dozens of kinds of additives, mineral fillers, and reinforcing fibers can be compounded into plastics. Bales of single-polymer plastics can be contaminated with pieces of incompatible polymers, such as plastic bottle caps or lids. For these reasons, 40–60 percent of recycled bottle PET ends up as fibers for clothes or carpet, and well over 50 percent of HDPE food/beverage bottles are turned into non-food containers, pipes, or plastic building materials.

Which is not to say recycling is impossible. Even some of the most difficult plastics can still be downcycled. The British company Recycling Technologies makes a machine that turns plastic waste into crude oil. Scotland's MacRebur is testing a recycled plastic road surface that is said to be stronger and more durable than asphalt roads and promises lower fuel consumption by reducing tire resistance on the road surface. The effect of plastic dust in the environment is another story, however.

Recycling Heroes

ore and more, companies are coming to realize that the fight against plastic pollution can be good for business.

Dell launched a pilot program in 2017 to recycle ocean plastics to make packaging trays for laptops, but it probably should have gone the extra mile and made the laptops themselves from that material. Who knows where the packaging trays will go.

Unilever and Nestlé have committed themselves to use 100 percent recyclable packaging. Adidas has begun making Ultraboost Uncaged Parley shoes from 100 percent recycled content.

For the past ten years, Jari Laine and Herbert Sixta at Aalto University and Kari Sinivuori at the University of Helsinki have been developing a way to recycle textiles, including polyester blends, using a nontoxic continuous-loop solvent called ionic liquid to dissolve old weaves into new via dry-jet wet spinning. Ioncell can turn used textiles, pulp, or even old newspapers into new biodegradable fabrics without chemicals. The fibers feel soft and are strong even when wet, retain their bright luster, and can be dyed like cotton.

David Katz founded the world's largest chain of stores taking only plastic barter. Everything in the store is purchased using plastic garbage, including school tuition, medical insurance, Wi-Fi, cell phone minutes, power, cooking fuel, and high-efficiency stoves. The stores are called the Plastic Bank. Any customer has the opportunity to earn a living by collecting material door to door, from the streets, from business to business, and taking it to the Plastic Bank, where it is weighed, checked for quality, and credited to the customer's online account, secure against robbery. Because it's a savings account, it becomes an asset that customers can borrow against. "It's no different than walking over acres of diamonds," Katz says. "If [you were] to walk over acres of diamonds but there was no store, no bank, no way to use the diamonds, no way to exchange them, they'd be worthless, too."

The Plastic Banks remove labels, caps, and grime, then either shred the plastic and pack it into bales for export or recover reusable bottles and other parts for reuse. They sell the baled plastics to manufacturers, such as Marks & Spencer and Henkel, who are branding it "social plastic."

Katz says, "Humans have produced over eight trillion kilograms of plastic, most of it still here as waste. Eight trillion kilograms. Worth roughly fifty cents a kilo, we're potentially unleashing a four-trillion-dollar value. See, I see social plastic as Bitcoin for the earth."

The Plastic Bank has a phone app that adds rewards, incentives, group prizes, and user ratings. "We've gamified recycling. We add fun and formality into an informal industry. We're operating in Haiti and the Philippines. We've selected staff and partners for Brazil. And this year, we're committing to India and Ethiopia. We're collecting hundreds and hundreds of tons of material. We continue to add partners and customers, and we increase our collection volumes every day. Now as a result of our program with Henkel, they've committed to use over one hundred million kilograms of material every year. That alone will put hundreds of millions of dollars into the hands of the poor in the emerging economies."

PLASTIC ARTISTS

In 2018, Beth Terry, author of *Plastic Free: How I Kicked the Plastic Habit and How You Can Too* and blogger at *My Plastic Free Life*, was contacted by Sreenivasulu M.R, a software professional in Bangalore, India, who had been making intricate miniatures of world-renowned architecture from used plastic-pen refills. His first model, Big Ben, took four days to construct. His largest and most complicated work to date, St. Philomena's Cathedral in Mysore, India, took eighteen months and required 2,500 pen refills. He's using these pieces to educate students about plastic.

Sreenivasulu M.R makes miniature structures out of pen refills.

To collect enough pen refills to create his projects, he visits schools and college campuses and sets up pen-refill collection boxes. He gives presentations called "Say No to Plastic." His informal survey revealed:

- Average number of pens used by a person in a year: 13
- Total number of refills the entire college used in a year: 18,200 (12 kg)
- Number of pen refills collected in a year: 1,580 (2 kg)
- Total number of refills used for making miniatures in a year: 1,200 (1.5 kg)
- Total number of refills not recycled: 17,000+ (10 kg)

Ghost Hunters

On one bright sunny morning, Healthy Seas, a global initiative that removes plastic rubbish from the oceans and recycles it into textile products, launched three boats—containing six divers from the Netherlands, two Italian divers, and three local fishermen—from a harbor on the island of Lipari. They were ghost hunters. For the past few days, project coordinator Veronika Mikos had been engaged in hauling nearly a ton of abandoned fishing gear from the seabed to shore. While diving, she had seen a ghost. Now she and her team were hunting it.

When they reached the spot where Mikos had seen the drift net, they dove until they found it, then bundled it with rope and attached a series of lift balloons. The years of debris caught in the net made it too heavy for the boat winch, so the fishermen brought it into the boat hand over hand. It weighed almost three tons, measured 650 feet long, and took an hour to drag aboard.

Healthy Seas cleaned and shipped the ghost net to a factory in Slovenia, where it became clothing. Aquafil, an Italian company sponsoring Mikos's mission, turned the plastic into Econyl, a regenerated nylon fiber, by chemically dissolving it into oil, extruding it to a resin, and spinning it into high-quality, high-performance yarn that will become bikinis, leisure apparel, and athletic jerseys.

Demand is high for fishing gear made from cheap, mixed-plastic polymers, and far more of those nets will be lost at sea than Healthy Seas and similar heroic efforts can recover.

"Microplastic" entered our lexicon in 2004 to describe the billions of minuscule bits of plastic that result from the breakdown of larger plastics like ghost nets. In 2015, a group led by the University of Georgia environmental engineer Jenna Jambeck estimated that between 4.8 and 12.7 million tons of plastic waste enter the ocean each year, a number she expects to double by 2025.

Around the UK, beaches saw a 30 percent increase in the abundance of large fragments (1–50 cm in size) and a 20 percent increase of smaller fragments (<1 cm) between 1998 and 2006. On the exponential curve, that is a doubling time of nineteen to twenty-eight years.

Ten years ago, the UK's Royal Society offered this warning:

> Plastics have transformed the surface of the planet, far beyond areas of human population density—fragments of all sizes are ubiquitous in soils to lake beds, from remote Antarctic island shores to tropical seabeds. Plastics turn up in bird nests, are worn by hermit crabs instead of shells, and are present in turtle stomachs. Humans generate considerable amounts of waste, and the quantities are increasing as standards of living and the population increase. Although quantities vary between countries, approximately 10 percent of solid waste is plastic. Up to 80 percent or sometimes more of the waste that accumulates on land, shorelines, the ocean surface, or seabed is plastic. The most common items are plastic films, such as carrier bags, which are easily windblown, as well as discarded fishing equipment and food and beverage packaging.
>
> From the first reports in the 1970s, it was only a few years before the widespread finding of plastic, including reports of microscopic fragments (20 μm in diameter). The abundance of microscopic fragments was greater in the 1980s and 1990s than in previous decades. It has also been suggested that plastic waste is deliberately being shredded into fragments to conceal and discarded at sea. Plastics of all sizes are now reaching the most remote and deepest parts of the planet, and although we have much better knowledge of their sources, quantities, and distribution, we still understand little about their longevity and effects on organisms. Further, we have made little progress in reducing the release of plastic to the environment . . . [and] our sustained demand for plastic means that contamination of the environment by micro-plastic pieces seems set to increase.

At the Factory Door

The light weight of plastic belies its relatively dense embodied energy and strong molecular bonds. The energy saved from recycling a

single plastic bottle can power a 100-watt light bulb for nearly an hour. Throwing that away is a waste of its qualities and the effort that went into its refining, as well as the deadly threat already outlined. The recycling rate for all plastics in the United States is meager—about 10 percent. Even plastic recycling's star performer—the PET water bottle—is well under 30 percent. In China, PET's recycling rate is 80 percent.

Changes in consumer demand can drive markets for recycling, as may be happening in Europe, but it's still moving in the opposite direction in North America. To get recycling infrastructure over a tipping point where it can become strong and stable may require regulatory intervention to reorient market forces, such as bottle taxes and restrictions on curbside trash collection. That led, rather than followed, changes in consumer demand in Europe.

Recycling is complicated by the quality of the material. Scraps that come from manufacturers are typically "clean," consistent in form and quality, and easily reprocessed, traded, and reused by the same industries that produce them. Postconsumer plastics are trickier. They get collected curbside or from bins; have to be separated from paper, metals, and other mixed municipal wastes; and then must be cleaned. Roughly only half of the plastics in the municipal recycling stream make their way from curbside collection to plastics recycling facilities. The rest are already too difficult to separate, handle, or clean.

Eliminated at the front gate of the facility or before are plastics for which no recycling infrastructure exists: plastic films that cannot be economically reclaimed; highly colored items; containers contaminated with milk, label adhesive, or motor oil; and polystyrene foam, because its low density makes the costs of storage and handling uneconomical. Painted or coated plastics can be prohibitively costly to clean. Of the plastics that are accepted by a facility, estimates range from 70–90 percent as the proportion that actually comes out of the reclamation process as usable recycled resin.

Once in the plant, cleaning can be as simple as separating the usual contents from a plastic diaper or as complicated as separating a cigarette butt from a plastic bottle. Repeated grinding, washing, and filtration help screen the incoming stream, and decontamination continues through and after the melting process and into extrusion. Then it remains to separate one grade of one specific kind and to try to determine whether it contains previous recycled content and how much. The polymers with previous recycled content are of a lower grade than the single-use variety and may not be reusable.

This stage uses high-speed sorting automation with density flotation separation or infrared spectra methods. The energy of reprocessing damages molecular properties further. Compensating for this additional "heat history," recycling plants add new stabilizers and enhancing additives that raise the cost, but the achievable purity will determine whether the recycled plastic can be used in new food and beverage packaging or will become pallets, strapping, or outdoor furniture. The downside is that many of these additives and stabilizers will render these products incapable of ever being recycled again.

Let me say that once more. Recycling plants are making single-use plastics, such as pallet strapping, that are incapable of ever being recycled again.

Well over 50 percent of HDPE food/beverage packaging is turned into non-food bottles, pipe, or plastic building materials. If a plastic can't be separated sufficiently even for those uses, another option is to break it down into its molecular constituents by thermolysis (chemical bath to depolymerize polymers into monomers or short-chain oligomers) or pyrolysis (high-temperature destruction). While chemical- and energy-intensive, these methods can at least produce usable petrochemicals or fuels.

Approaching from the Opposite End

At the Center for Sustainable Polymers, scientists are inserting poison pills into polymers by capping the ends of the long chains or linking the chains together into networks designed to fail if they encounter environmental triggers, such as light or seawater. In theory, the long chains unzip on their own once removed from the conditions that made them endure. This gives self-unzipping polymers an edge over biodegradable ones, says Elizabeth Gillies, a polymer chemist at Western University in London, Ontario. "We

can have a big change in properties or complete degradation of the polymer just from one event."

One problem with this approach is that consumers may not like watching a milk container begin leaking all over the kitchen counter because someone spilled some salt on a wet surface. Zippers have to be designed with as much care as the products they unzip.

It is unlikely that a 400 degrees C trigger would be encountered anywhere in the average home, but if you are a recycler collecting used Huggies to unzip at that temperature, you may want to ask where you will find and how you will pay for that much heat for tens of millions of them, not to mention what you will do with any unwanted contaminants they picked up during normal use.

Systemic Change

Not so long ago, pretty much all baby food came in jars, pet food came in cans and boxes, and frozen vegetables and trail mix came in bags—and you could recycle that packaging. Now those products are often sold in convenient, resealable plastic pouches that generally can't be recycled.

Stand-up pouches are expected to become a $40.65 billion market by 2022. Layfield Flexible Packaging, a Canadian company, makes pouches for medical uses, such as blood bags, as well as food and consumer packaging. Company president Mark Rose says there are good reasons—including environmental ones—why pouches have come to replace so many aluminum cans, glass jars, and hard plastic containers. They're lighter and use less material, saving energy and costs of manufacturing and transportation. They're made by layering different kinds of plastics, each with unique advantages. An outer polyester holds ink colors well and keeps oxygen out, so the food stays fresher. A middle laminate of polyethylene keeps out moisture. An inner, metal-coated plastic layer keeps out light. And the packaging is rigid enough to stand up on a shelf, where they can take up less space than jars or cans. The downside? Because they're made of layers of adhesive and different plastics with different melting temperatures, they're not recyclable. Moreover, in most recycling sorting facilities, flat objects, such as empty stand-up pouches, end up in the paper stream and get burned or landfilled.

Some pouches can be recycled without much processing into outdoor furniture, decking, and fence posts. What can't be recovered can be turned into energy pellets that will be burned as a replacement for coal or diesel at industrial sites like cement kilns.

While Mark Rose is convinced that pouches are still more environmentally friendly than other kinds of packaging, he has spent years trying to develop versions that are greener to dispose of. In 2018, Layfield launched two recyclable types of pouches, each with only one kind of plastic—either polyethylene or polypropylene. Because they are not as good at stopping oxygen or providing transparent windows on the side, they are mostly used to hold things that don't spoil.

One company, TerraCycle, offers free recycling programs funded by brands, manufacturers, and retailers around the world to help you collect and recycle your hard-to-recycle waste, such as stand-up pouches. The brainchild of Princeton University student Tom Szaky, TerraCycle has grown from 2001 into the global leader in collecting and repurposing hard-to-recycle waste, operating in over twenty countries, engaging over eighty million people, and recycling billions of pieces of waste through various platforms.

Still, since most pouches look the same, most are being rejected by recycling programs. Jamie Rhodes, program director for Upstream, a US nonprofit based in Rhode Island that ran a campaign against stand-up pouches, says the onus should be on the manufacturers marketing the packaging as recyclable to ensure it is actually recycled. "If they're making the claim this is a recyclable pouch, but they're not willing to invest in the system to actually collect and recycle it, then their claim is just greenwashing."

Layfield's Rose agrees and acknowledges it's hard to efficiently collect stand-up pouches, given how light they are and the fact that recyclables are sold by the ton. "It may not be environmentally advantageous to recycle it," he said. "I still think, at some point in time, a stand-up pouch is going to end up in a landfill."

Because of that, Layfield has developed a special kind of plastic called Bio-Flex. Pouches made with Bio-Flex are designed to decompose in a landfill within ten years, although they emit the dangerous greenhouse gas methane as they decay. So that's only part of the solution, according to Rhodes, who thinks people must also reduce their reliance on single-use plastics, especially when alternatives are available.

"I don't think we can recycle our way out of this problem."

Currently, half the plastics we use are thrown away after a single use. That includes 2.5 million plastic bottles and 2.9 million Styrofoam cups, just in the United States. That's every hour. Every year, enough plastics are thrown away to circle the world four times. But then, even the plastics we recycle have been circling the globe.

The China Crisis

For many years, much of the plastic garbage generated in the United States and Europe went halfway around the world, aboard container ships to China. In 2017, China shocked the world by saying it had had enough. It wanted no more.

Specifically no more plastic. And even a tiny bit of plastic (0.5 percent) found in a two-ton bale of paper received for recycling in Shenzhen or Shanghai meant that the whole bale would be marked "contaminated" and rejected. To avoid extra shipping, China now sends its officials to recycling centers in the United States to inspect bales before they leave the warehouse.

Over the past thirty years, while China was filling container ships with recyclable garbage to bring home and feed its factories, consumers in North America and Europe got lazy. Ordinary bottles, cans, and paper went into blue bins and were transported to a local single-stream recycling facility. It was up to that facility to sort it down further and then sell it to China. As long as that lasted, it was profitable for the recycling industry. But vastly more recycling, with less attention to detail, resulted in an increasingly adulterated product leaving wealthy countries, with poorer China bearing the cost of cleaning. Mixed media, such as plastic/paper pouches and foil-lined drink cartons, could be anywhere in the stream.

Finally, China had had enough. It was losing money, losing its environment, and harming its people's health. President Xi Jinping let it be known the free lunch had ended.

The United States, with less than 4 percent of the world's population, generates one-third of the world's waste. In 2018, municipalities that had been getting rid of their recyclables for free, or even getting paid, started having to figure out what they were now going to do with it all.

CHAPTER 5

Fantastic Plastic

Never underestimate the cleverness of mushrooms to find new food.

PAUL STAMETS

One billion years before the introduction of plants, fungi thrived on the land in forests of tall mushroom trees. All animals, humans included, are a later evolutionary branch extending from those trees. Fungi are today the primary decomposers of matter, recycling nutrients, medicines, and water to plants and endowing fresh soil with new growth.

The wisdom of these ancient organisms holds solutions to many of the problems we confront with plastics. Recycling is often thought to be the best option for dealing with plastics, but as we have seen, it doesn't represent a complete solution. Eight percent of plastics are thermosets, which means that they can't be remolded or recycled. Some elastomers and thermoplastics can be recycled but require careful sorting, yield a lower-grade product, and often come at a higher cost. Biodegradable plastics may not degrade very quickly, can gum up the works, and raise questions of food versus fuel or the need for forests in land and climate management.

Enter mushrooms. Through the refinements wrought over a billion years of coevolution, fungi have diversified to fill several roles in nature. As mycologist Paul Stamets says, "Fungi are the interface organisms between life and death." The decomposing, or saprotrophic (living on dead matter), mycorrhizae have developed powerful enzymes to break down long-chain hydrocarbons. While animals were evolving internal stomachs and intestines, fungi chose to keep their potent digestive acids external to their bodies. It is similar to the way a common housefly eats. Fungi dissolve their food first and then ingest it as a liquid. Fungi can even break down bare rock to withdraw minerals. They create new life from inanimate objects.

More than a century ago, when plastics first began appearing, manufacturers tested their products to determine longevity in comparison to ceramics, wood, metal, and glass. One test was simply to bury pieces of plastic in the ground and dig them up a year or two later. Experimenters found fungi and bacteria deconstructing the polymers. The digestive enzymes the fungi produced were so refined and powerful that they had made the plastic available as food for the bacteria. This led to antifungal or antibacterial additives designed to thwart natural decay.

In general, fungi are more adept at disassembly than bacteria. Research has identified various species and genera best suited to this work, but until recently little progress was made toward finding or breeding the best strains of natural decomposers.

At Yale University, undergraduate students extracted enzymes from an endophytic fungus, *Pestalotiopsis microspora*, drawn from an Amazonian plant. Endophytic fungi are not soil-dwelling like many fungi but live inside or between a plant's cell walls. When the team applied enzymes harvested from the endophyte to a sample of polyurethane, the endophyte was able to survive off the plastic digestate as its only food source. Perhaps more importantly, it could perform this function both in the presence and absence of oxygen (aerobically and anaerobically). Most ground-dwelling fungi can't survive in the anaerobic environment of landfills.

Endophytic fungi are possibly one of the most diverse categories of fungi, with any given plant potentially containing hundreds of species. The number of endophytes in the world is unknown, but it would be reasonable to speculate that many hold even more robust capabilities. New clues may be found in our newfound ability to map fungal genomes to understand family trees. It is likely, for instance, that fungi whose ancestors were once living as endophytes within lichens but are now free-roaming saprophytes could have the right saliva to attack even the most decay-resistant polymers.

It may not even be necessary to discover better and better decomposing candidates through time-consuming trial and error. What we are learning from epigenetics suggests it might be easier to simply "train" a given fungus to consume food sources that it wouldn't previously have chosen. In remediation work with chemicals, for example, the targeted pollutant is introduced to a fungus at ever-increasing concentrations until the fungus learns to produce the right enzyme at the right amount to be able to survive and thrive in what were previously toxic conditions. It does this by flipping genetic switches within cells to retask the production lines from one enzyme to another while building its own resistance to the toxicity.

It may be that some plastics will require not a single fungal specialist but several. When a plant dies and drops to the ground, there is a progression of bacterial and fungal decomposers, worms, and microbes that work in combination, or sequence, to complete the entire breakdown and recycling process so that all nutrients are efficiently recaptured. Primary, secondary, and tertiary fungi all play a role. A similar progression may be required to break down plastics. Peter McCoy, author of *Radical Mycology,* says, "Such considerate and non-reductionist approaches that reflect and respect the complexity of nature will likely be the key to begin dealing with the plastics problem."

As part of its 2018 initiative to cut down on the use of single-use items where feasible, or to replace them with more sustainable alternatives, the Parliament of the UK rolled out a new range of bio-based, certified-compostable catering items, such as coffee cups, soup containers, and salad boxes, in the House of Commons and the House of Lords. The government plans to stop providing bottled water in the Parliament, following the example of Buckingham Palace. New waste bins will be installed in Parliament and the royal palaces to capture used compostable items, and an organic recycling facility has agreed to make Her Majesty's compost.

Hasso von Pogrell, the managing director of European Bioplastics, said, "The concepts of prevention, re-use, single-use, and sound waste management need to be optimally combined to make sustainability and circularity happen." Is it possible that the plastics of the future will be made from materials designed to be naturally decomposed? This is a question being posed at business and science conferences convened to deal with the problem of the indestructible and ubiquitous character of plastics. What we had once thought was our best friend has turned into a daunting foe. World leaders are asking: could there be something we should have done differently and can we still do that now?

Lest we forget, the first plastic was a bioplastic from nitrocellulose (see page 7), made by Alexander Parkes in 1862 for the Great London Exhibition. German chemists in 1897 made Galalith, a milk-based bioplastic still produced today for buttons. The first fossil-fuel polymer was Bakelite (see page 8) in 1907, made from formaldehyde and coal tar, but it was the exception. In 1912, Jacques E. Brandenberger invented cellophane out of wood, cotton, or hemp cellulose. In the 1920s, Wallace Carothers made polylactic acid (PLA; see page 25) from corn dextrose. In 1926, Maurice Lemoigne made polyhydroxybutyrate (PHB; see page 32) from bacteria. The first polymeric automobile was made from soybeans by Henry Ford.

In the 1950s, bioplastics entered a long decline thanks to cheap and abundant oil, coal, and gas, but the development of biosynthetic plastics continued. In 1983, Marlborough Biopolymers developed a bacteria-based bioplastic called biopal. In the late 1990s, the Biotec company introduced its Bioflex into extruded films and injection molding lines. Today those products, or something very similar, are in sacks, bags, trash bags, mulch foils, hygiene products, diaper films, bubble wraps, protective clothing, gloves, double-rib bags, labels, barrier ribbons, trays, flowerpots, freezer products and packaging, cups, pharmaceutical packaging, disposable cutlery, cans, containers, pre-formed pieces, CD trays, cemetery articles, golf tees, toys, and writing materials.

Bioplastics today fall into either or both of two broad categories:

1. Bio-based plastics that are derived from renewable resources
2. Biodegradable plastics that meet standards for biodegradability and compostability

Biodegradable plastics can be petroleum-based, and bio-based plastics can be either biodegradable or non-biodegradable.

Along with bioplastics that are inherently biodegradable, some plastics are being formulated with oxo-biodegradability additives to speed decomposition in ambient air at the end of the product's expected life. For some of these, the first step is photodegradation (breakdown by sunlight) followed by biodegradation when exposed to microorganisms.

Any true biodegradation must happen in a reasonable amount of time—weeks or months—rather than the dozens or hundreds of years required for persistent fungi and bacteria to degrade traditional plastics. Ideally the process should happen in nearly any natural environment or in a landfill, but in reality, it often requires controlled composting conditions for the proper mixture of air, heat, and microbes.

Because of the fragmentation in the market and ambiguous definitions, it is difficult to describe the total market size for bioplastics, but some estimates put global production capacity at 325 to 350 thousand tons in 2018. By contrast, global production of polyethylene (PE) and polypropylene (PP) are estimated at over 150 *million* tons. A study by the European Union found that demand for bioplastics, just in Europe, already exceeds two million tons per year, six times the available world supply. Today bioplastics represent less than 1 percent of the overall plastics market, but that is about to change.

The most environmentally minded bioplastic maker today could be Gunter Pauli's Novamont SpA, a joint investment of Intesa Sanpaolo and Investitori Associati. Starting as a research center, Novamont has more than two hundred employees and reinvests its sales profits in research and development. Thirty percent of its workforce is in R&D. Based in Novara, Italy, with production sites in Sardinia, its core model is a "bio-refinery model closely linked to the territory" using only renewable raw materials of agricultural origin. But despite winning numerous awards for its ecological, bioregional, and human betterment philosophy, Novamont faces some stiff competition.

Monsanto acquired Biopol, which makes plant polymers, from the Swiss pharmaceutical company AstraZeneca. It then sold the bioplastics brand to Metabolix, Inc. In 2007 Metabolix market-tested its first 100 percent biodegradable plastic, called Mirel, made from corn sugar fermented by genetically modified organisms (GMOs) and processed into PLA. Metabolix, now called Yield10 Bioscience, then formed a joint venture with transnational grain giant Archer Daniels Midland called Telles and produced Mirel for Target stores for the paper in gift cards. Target's tests proved the cards degraded by composting in about 120 days. Mirel was then chosen by Paper Mate for its line of biodegradable pens. Meanwhile, Cargill and Dow Chemical rebranded themselves as NatureWorks to become the leading GMO-PLA producer. These deep-pocket companies are moving into the bioplastics space by employing biotech without the competitive baggage of the strict environmental ethics a GMO approach requires.

It's the Biology, Stupid

Wax worms feed on the wax in beehives, and so beekeepers typically remove them. Federica Bertocchini, a scientist with the Spanish National Research Council, was deworming her beehives when she noticed that the worms had chewed holes through the plastic bag she was collecting them in. She created a paste from the wax worms in a blender and spread the worm homogenate on plastic. The results showed that something being excreted by the worms degraded the plastic at a molecular level, but Bertocchini and her colleagues couldn't tell if it was an enzyme produced by the worms, a bacteria found in the worm, or a combination of both. What they could confirm was that the polymer backbone was breaking down into smaller monomers: PE plastic became ethylene glycol.

A 2018 study of the gut microbiome of meal moths (the larvae of *Plodia interpunctella*) found that the worms could adjust their gut microflora to digest the unique polymer structures of chemically dissimilar plastics. Even plastics such as PE and PS that were previously believed to be nonbiodegradable were degraded by the strains of bacteria living in the stomachs of the worms.

In 2016, Japanese scientists exploring an old waste dump discovered a bacterium that had naturally evolved to eat plastic. After two years of laboratory work, an international team revealed the detailed structure of the enzyme produced by the microbe that did the work. In the process of tweaking the enzyme to see how it had evolved, they inadvertently made a new substance even better at breaking down the PET in soft drink bottles.

"What actually turned out was we improved the enzyme, which was a bit of a shock," said professor John McGeehan at the University of Portsmouth, UK, who led the research. "What we are hoping to do is use this enzyme to turn this plastic back into its original components, so we can literally recycle it back to plastic." The goal is to recycle clear plastic bottles back into clear plastic bottles.

The research began by determining the precise structure of the enzyme produced by the Japanese microbe. The team used an intense beam of X-rays that is ten billion times brighter than the sun and can reveal individual atoms. What happened when they put the enzyme under the ray was unexpected but almost straight from a science-fiction movie. They accidentally mutated its ability to eat PET.

"It is a modest improvement—20 percent better—but that is not the point," said McGeehan. "It's incredible because it tells us that the enzyme is not yet optimized. It gives us scope to use all the technology used in other enzyme development for years and years and make a super-fast enzyme."

Of course, genetically modifying plastic-eating bacteria carries its own risks, as soil scientist Elaine Ingham discovered when working as a researcher at Oregon State University in the early 1990s. She later wrote:

> We started testing the ecological impacts of most of the genetically engineered organisms being produced at that time. The question our lab was asked to address was: Did these engineered organisms have any impact out there in the real world? The first fourteen species that we worked on—microorganisms, bacteria, and fungi—were organisms incapable of surviving in the natural environment. Putting them in the world would be like taking penguins from the South Pole and dropping them into the La Brea tar pits. Would there be any ecological effect if we dropped a penguin into the middle of the tar pit? Probably not; the impact would be rapidly absorbed by the system.

On this basis, the USDA Animal and Plant Health Inspection Service, the regulatory agency that was determining US policy on genetically engineered organisms, gave approval to the research. Ingham describes what happened next:

> GMO number fifteen, however, was a very different story. *Klebsiella planticola*, the bacterium that is the parent organism of this new strain, lives in soils everywhere. It's one of the few truly universal species of bacteria, growing in the root systems of all plants and decomposing plant litter in every ecosystem in the world.
>
> The genetic engineers took genetic material from another bacterium and inserted that trait in the GMO to allow *Klebsiella planticola* to produce alcohol. The aim of this genetic modification was to eliminate the burning of farm fields to rid them of plant matter after harvest. The idea was that you could, instead, rake up all that plant residue, put it in a bucket, and inoculate it with the engineered bacterium, and in about two weeks, you would have a material that contained about 17 percent alcohol. The alcohol could be extracted and used for gasohol, for cleaning windows, or for myriad other uses: cooking with alcohol in Third World countries, for instance.
>
> The genetic engineers thought this transformation would bring huge benefits. We would no longer have to burn fields, we would breathe better in the fall, and both the company and farmers would get a product that could be sold. There was actually a fourth win: the sludge at the bottom of the bucket is an organic fertilizer, and there are no waste products from that material. So what's the problem?

The problem was that there was nothing to turn the GMO off once it was loose in the environment.

> Once it's there and growing in the root systems of your plants, it's producing alcohol. What level of alcohol is toxic to plants? It's one part per million. How much alcohol does this engineered organism produce? Seventeen parts per million. Very soon you will have drunk dead plants.

At the 1995 Madrid meeting of the United Nations biosafety protocol study group, Ingham related the story of *Klebsiella planticola* as an example of the lack of adequate testing for the ecological impact of genetically engineered organisms. The biotech companies objected that it costs too much to do this kind of environmental testing. Ingham said it should be worth any amount of money to do these simple experiments because if the wrong bacterium were let loose in the environment, it could cause very significant problems for our food supply.

How far does a single-point inoculation of a genetically engineered organism spread in one year? Ingham explains:

> An engineered *Rhizobium* bacterium that was released in Louisiana in the mid-1990s spread eleven miles per year and has by now dispersed across the North American continent.
>
> At these United Nations meetings, I warned that corn pollen is going to move a lot more than three feet away from the plant. "Oh no," said the biotechnology representatives present. "Corn pollen falls out of the air three feet from the plant."
>
> I would say, "Wait a minute, you've never heard of bees? How about birds? And insects? And wind?"
>
> "Oh no, it falls out of the air within three feet of the plant."
>
> Why do our bureaucrats choose to believe these "scientists"? Just open any plant textbook and you find that corn pollen can be found in the Antarctic and the Arctic. But if you listen to Monsanto, corn pollen can't possibly be there.

We should not even consider releasing a genetically modified microbe that can eat plastic into the wild. Do you remember the parent bacterium the Japanese scientists found in that landfill? It had evolved from bacteria that could slowly decompose cutin, a natural polymer used as a protective coating by plants. Putting that process on steroids—speeding it up a thousandfold or a millionfold—hazards all plants with hard cutin coatings.

Evolving Feedstocks

First-generation bioplastics were made from food crops, including corn, wheat, sugarcane, potato, sugar beet, rice, and plant oil. In other words, they used food otherwise destined for human or animal consumption. Second-generation feedstocks were either nonfood crops (wood; short-rotation woody crops, such as poplar, willow, miscanthus; or switchgrass) or waste materials from food crops (waste vegetable oil, sugarcane bagasse, palm fruit bunches, corncobs, and wheat, oat, or rice straw).

Third-generation feedstock is biomass derived from algae. Algae don't need fertilizers, pesticides, herbicides, or farmland. Some algal plastics biodegrade within twelve weeks in soil and five hours in water.

By 2012, bioplastics had been synthesized from vegetable waste (parsley and spinach stems), cacao pods, rice hulls, and seaweed. In 2013, a patent was awarded for a bioplastic derived from blood and intended for use in grafting cartilage, tendons, ligaments, and bones, and for delivering stem cells to sites in the body where they can grow out into new tissue. In 2016, one-upping Henry Ford, scientists made a car bumper using plastic derived from banana peels, which withstood crash-test-dummy collisions into concrete barriers.

None of the currently available bioplastics—which can be considered first-generation products—require the use of genetically modified crops, although GMO corn is a standard feedstock where it is still legal, such as in the US. There are ominous signs ahead, however, because of the willingness of some of the larger agribusiness firms to gamble with genetic manipulation and sometimes play fast and loose with safety. The second-generation bioplastics manufacturing technologies under development employ the "plant factory" model, using genetically modified crops or genetically modified bacteria to optimize efficiency.

Standards for Biodegradability

Showing that a material is genuinely biodegradable first requires testing it against an accepted standard, but there are a confusing number of accepted ISO, EN, and ASTM standards for measuring or making claims of biodegradability. Six major certification systems exist worldwide concerning compostability: ABA (Australia); BPI (USA); DIN CERTCO, Vinçotte, and European Bioplastics (Europe); and JBPA (Japan). These systems are all based on the same international standards but have distinct and nuanced criteria for certification.

Standards are important, because unless a piece of plastic is converted by microbes quickly into benign short-chain molecules, makers shouldn't be allowed to make a claim of biodegradability.

ASTM International, formerly known as American Society for Testing and Materials, is an international standards organization that develops and publishes voluntary consensus technical standards for a wide range of materials, products, systems, and services. At this writing, 12,575 ASTM voluntary consensus standards operate globally. Standard D6954 is a three-tiered testing guide for measuring the degradation, biodegradation, and ecological impact of oxo-degradable plastic.

So what happens if a bioplastic maker introduces a product that does not meet the ASTM D6954? First, any claims of biodegradability would be misleading at best and harmful at worst. Maybe the plastic does break down, but only partially. Disassociated from their strongly bonded pendant chains, those partly biodegraded backbones can attract and concentrate toxins already in the environment, such as glyphosates or nicotinoids. If these new polymers are then eaten by mice, birds, or fish, the toxins can wreak havoc on wildlife and ecosystems before migrating into the human food chain.

Whether a product can be certified to the ASTM D6954 standard would likely vary depending on whether the application was made in Australia, the United States, Europe, or Japan. If that is confusing for users, it's even more discouraging for manufacturers. When the standards are that varied and complex, the tendency is to ignore or abuse them, and as we can see, this is very dangerous.

Common Bioplastics

hese are some common bioplastics, with varying degrees of biodegradability:

Starch is cheap, abundant, and renewable and makes up about half of the bioplastics market. It is so simple it can be made at home, and if you ever made paper-mâché as a child, chances are you have already made some. Flexibilizers and plasticizers, such as sorbitol and glycerine, can also be added for a resulting bioplastic called thermoplastic starch. Starch-based bioplastics can be blended to produce starch/polylactic acid, which has better water-shedding and mechanical properties. These blends are also compostable, but others, such as the starch/polyolefin blends, are not, although they do have a lower carbon footprint than petroleum-based plastic.

Cellulose is the original natural plastic and still the most abundant organic polymer on Earth. It's the primary structural ingredient of the cell walls of green plants, many forms of algae, and the oomycetes (water molds). Some species of bacteria secrete it to form biofilms. The cellulose content of cotton fiber is 90 percent, wood 40–50 percent, and hemp approximately 57 percent. Cellulosic fibers added to starches can improve mechanical properties, permeability to gas, and water resistance due to being less hydrophilic (water loving) than starch.

Wheat gluten, soy protein, and casein (from milk) show promising properties as biodegradable polymers. Soy proteins have been used in plastic production for over one hundred years but lost out to fossil-based plastics due to their water sensitivity and relatively high cost. A new frontier lies in blends of soy with biodegradable polyesters to boost water resistance and lower cost.

Polylactic acid (PLA) is a transparent plastic produced from corn or dextrose. It is superficially similar to polystyrene but degrades to nontoxic products. Unfortunately, it is inferior to polystyrene in impact strength, heat and cold tolerance, and blocking air permeability compared to other blends in commercial use.

Poly-3-hydroxybutyrate (PHB) is a polyester produced by certain bacteria processing glucose, cornstarch, or wastewater. It's similar to polypropylene. Brazil is taking it to large-scale production using the waste by-products of growing sugar. It can be processed into a transparent film with a melting point higher than 130 degrees C and is 100 percent biodegradable without residue.

Polyhydroxyalkanoates (PHA) are produced in nature by bacterial fermentation of sugar or lipids. They are produced by the bacteria to store carbon and energy. More than 150 different monomers can be combined within this family to make materials with widely different properties. PHA is more ductile and less elastic than other plastics and is also biodegradable. PHA is widely used in the medical industry.

Polyamide 11 (PA 11) is known under the trade name Rilsan and comes from castor oil. Because of its toughness, flexibility, and chemical and permeation resistance, it can be found in automotive fuel lines, pneumatic air brake tubing, electrical cable anti-termite sheathing, flexible oil and gas pipes, control fluid umbilicals, sports shoes, and catheters. It is not biodegradable. A similar plastic is **Polyamide 410** (PA 410), derived 70 percent from castor oil, using the trade name EcoPaXX and sold in several different blends.

Bio-derived polyethylene is chemically and physically identical to traditional polyethylene but produced by fermentation of sugarcane, beets, or corn. It does not biodegrade but can be recycled. The Brazilian chemicals group Braskem claims that, using its method of producing sugarcane and refining polymer, it captures (removes from the atmosphere) 2.15 tons of CO_2 per ton of Green Polyethylene it produces. That is a big deal.

Polyhydroxyurethane (PHU) is bio-based and isocyanate-free polyurethane. The substitution of natural-oil-based polyols was the first route developed. The second strategy is blending polyamines to polycyclic carbonates without the use of harmful isocyanate. The chemical transformation of epoxidized vegetable oils or glycerine-carbonate-based intermediates provided that path. New families of bio-based PHUs are capable of recycling and reprocessing.

Polyurethanes, polyesters, epoxy resins, and other types of polymers have been developed with comparable properties to petroplastics through the development of olefin metathesis from vegetable oils and microalgae. Algae has been called "the ultimate winner" in the bio-feedstock race because it absorbs water, sunlight, and CO_2 from the air, the organisms grow rapidly, and their production does not disrupt food markets. Bio-oil production rate per unit of cultivation area is fifteen times higher than with other biomass sources.

Surplus soybean oil, reinforced with lignin from wood, is being developed by the United Soybean Board into resin monomers for sheet and bulk molding compounds and resin transfer molding, including a dashboard panel for John Deere tractors. Ground-up corncobs are being developed into a biofiller for a vinyl window and door manufacturer. Non-edible cellulose and cashew nutshell oil (cardanol) have been chemically bonded with other additives to create a 70 percent plant-based thermoplastic material said to have multiple times the strength and heat resistance of PLA and the equivalent water resistance.

Keratin resin made from poultry feathers, combined with a poly-olefin, is being molded into biodegradable flowerpots for nurseries, making use of at least some of the three billion pounds of chicken feathers created by US poultry processors per year (roughly 80 percent is currently sent to landfills).

Resins are also being experimentally produced from carbon dioxide itself and synthesized into a polypropylene carbonate resin (PPC). The latest is a 44 percent CO_2-based Greenpol PPC using a proprietary catalyst and a continuous polymerization process. Its makers see potential uses in packaging materials, supplanting petroplastic poly-olefins. Because the UN's Nobel Prize–winning science advisory, the Intergovernmental Panel on Climate Change (IPCC), has reported that carbon dioxide removal will be essential if catastrophic global warming is to be avoided in this century, much of our future energy will likely derive from biomass energy with carbon capture and storage (BECCS). Greenpol or PPC resins like it could supply the missing CCS part of that equation at a profit, rather than attempting to pump CO_2 liquids into deep reservoirs or to the ocean floor.

In addition to bioplastics, there are many more novel or experimental polymers being studied or test-marketed. One is high-performance nanoporous nanofibers from biodegradable clays, which form nano-fiber webs. Early researchers report that these nanopolymers form physically and chemically responsive functional groups that "tend to strong self-assembly." When stressed, the increased loading of partner polymer nanocomposites speeds self-adaptation, which improves the morphology and thermal behaviors of these webs, making them potentially useful in nanomedicine, nanomicrobiology, and water filtration.

Self-assembling nanofiber webs sounds like the future has just arrived.

CHAPTER 6

The Fabrics of Society

> We know to avoid plastics because they are made of non-renewable fossil fuels, they are not biodegradable, and they leach hormones and toxic chemicals. What many of us are unaware of is that plastics make up the fabric of our everyday life. Look down at your shoes, socks, and pants. Do you know what the fabric composition of your clothes is? Look at your rug, your couch, your bed, and the sheets on it. Do you know what they're made out of? Chances are they're plastic, or at least part plastic.
>
> **CAMILLE SCHEIDT**
> *UNRAVELING THREADS: HOW TO HAVE A SUSTAINABLE WARDROBE IN THE AGE OF PLASTIC FABRIC*

Synthetic clothing, including polyester, polyamide, nylon, and acrylic, is very cheap to make and very bad for you and other living things. Because of its low price tag, it is tempting to buy, and retailers and manufacturers may even make it hard for you to choose otherwise. They hide plastic microfibers in budget-friendly fabrics called blends. "China silk" is a term in the textile industry for a 100 percent polyester fiber woven to resemble the sheen and durability of insect-derived silk.

Synthetic fabrics, such as nylon and polyester, shed thousands of microscopic fibers with each wash cycle. Once scientists started showing how these fibers end up on your dinner plate after passing through little fish to bigger fish, newspapers ran articles with headlines such as "Yoga pants are destroying the Earth." Seizing the moment, eco-conscious brands began selling a washing-machine pellet that they claim catches "some" of the plastic sloughing off clothing. (Patagonia calls theirs "Guppyfriend.")

Stephen Buranyi, writing for *The Guardian*, lamented:

> It slips through our fingers and our water filters and sloshes into rivers and oceans like effluent from a sinister industrial factory. It is no longer embodied by a Big Mac container on the side of the road. It has come to seem more like a previously unnoticed chemical listed halfway down

> the small print on a hairspray bottle, ready to mutate fish or punch a hole in the ozone layer.

One-third of fish caught in the North Atlantic are contaminated with microplastic. It is even found in benthic animals living thousands of meters below the sea surface. Eighty-three percent of drinking water samples from around the world are contaminated with plastic fibers. Much of this contamination of fresh and saltwater comes when synthetic-fiber-based clothing is worn and washed.

It won't help you if you decide that rather than throw your clothes in the washing machine you will take them all to be dry-cleaned. The most common dry-cleaning solvent is PCE (perchloroethylene—"perc"). As Camille Scheidt reveals, "There are no perks to perc."

> Once the solvent vaporizes, it is easily inhaled. Because of this, both dry cleaning employees and customers are directly at risk of breathing in the chemical. The dangers of perc are not isolated to the dry cleaning facility. Perc can follow you home. The chemical remains in dry-cleaned clothing long after it leaves the cleaner, and the levels of perc in the garment will accumulate with each cleaning process. But, as you just learned, the perc doesn't just stay in your clothing; it off-gases. A study found that if you were to put four freshly dry-cleaned sweaters in your car and step into the grocery store for an hour on a warm day, you would return to a car that was well exceeding the safe limit of perc exposure.
>
> But perc pollution reaches much further than your home and car. The contaminant has been detected in groundwater and both public and private wells. It's also found in soil. Perc can become airborne from soil and water, and once in the air can be inhaled. The effect of perc on our bodies is severe. Short-term exposure at low levels can cause inebriation, dizziness, and irritation in the eyes, nose, mouth, throat, and respiratory tract. Short-term exposure to perc at high levels can cause fluid buildup in the lungs, difficulty speaking and walking, headaches, drowsiness, dizziness, nausea, and irritation of skin and the respiratory system. If a person's exposure to perc is at a high level, even for a short time, the chemical can cause unconsciousness and death.
>
> Prolonged exposure to perc can result in damage to the central nervous system, kidneys, and liver. It is recognized as a probable human carcinogen and linked to cases of cervical cancer, bladder cancer, esophageal cancer, kidney cancer, lung cancer, breast cancer, and non-Hodgkin's lymphoma. In 2007, it was estimated that one in ten wells in California was contaminated with perc.

Fashion, unlike many other aspects of the plastic problem, is something consumers can change both thoroughly and rapidly. Besides producing and buying fabrics that last longer and can be recycled, we can purchase clothing made from organically produced materials that naturally biodegrade, such as cotton, silk, linen, and wool. We can wash only when the clothes, especially outerwear, absolutely require it. We also have to be aware, when we are buying, not to purchase blends. Many fabrics can be recycled, even acrylics, but if it requires the entire structure to be disassembled, thread by thread, remanufacturers may shy away.

Petroplastic fabrics are something we can, and must, refuse. Surely we can replace these with safer, healthier, bio-based and biodegradable natural analogs. Finland's bioeconomy has expanded on the strength of its forests. New developments in the fabric industry there will extend the value chain of forest biomass to include cellulose-based nonwoven textiles, estimated to reach 47.7 billion euros in 2020.

"Cellulose fibers can be utilized in all textiles that can replace cotton and viscose which both have sustainability issues related to their production. Government strategy in Finland aims to double the current bioeconomy turnover from 60 billion euros to 100 billion euros before 2025," says Tuula Savola, program manager of Business Finland's Bio-Nets program.

On the horizon in the rest of the world are new fabrics that provide a better experience and range of qualities than toxic and indestructible synthetics and blends. These include sustainably harvested cork fabric as an alternative to leather; fish skin; mushroom "skin" (the cap skin from *Phellinus ellipsoideus*, native to subtropical forests); pellemela, sustainably sourced from discarded apple peels and core waste from juiced apples; Piñatex from pineapple processing waste; Orange Fiber yarn and silks; TENCEL from beech and eucalyptus; and hemp, which is UV-, mold-, and mildew-resistant, naturally antimicrobial, absorbent, and durable.

Confronting Plastic Culture

Plastic culture is another matter. At the end of 2018, the Club of Rome, a fifty-year-old think tank known for accurately predicting the future, issued this warning:

> The prevailing mantra that all economic growth is good defies the reality of life on a finite planet with finite resources. There is an urgent need for new economic thinking and new indicators that value quality as well as quantity in our economic metrics.

Since most of the qualities we seek in plastic products can now be found in biodegradable bioplastics, what we most need to do is to change the economic metrics—how value is assigned. In many ways that realignment dovetails neatly with what is required to arrest and reverse climate change. Here are portions of the roadmap proposed by the Club of Rome, which I have amended slightly to include plastics:

- Introduce realistic pricing and taxation to reflect the actual cost of fossil fuel use and embedded carbon.
- Introduce carbon (or non-green plastic) floor prices.
- Tax embedded carbon (or non-green plastic) through targeted sales taxes.
- Fund research, development, and innovation.
- Converge carbon (and green plastic) markets and instruments into a worldwide structure.
- Replace GDP growth as the primary objective for societal progress.
- Adopt new indicators—such as the Genuine Progress Indicator (GPI)—that accurately measure human development, welfare, and well-being rather than production growth.
- Establish explicit funding and retraining programs for displaced workers and communities.
- Provide government assistance to enable older industries to diversify to lower-carbon (and green plastic) production.
- Reframe business models and roles for declining industries such as oil, gas, and coal.
- Create an international convention, applying to nations and non-state actors alike, with legally enforceable rules and mechanisms for policing the global commons.
- Support citizen action and litigation against countries and actors exceeding legal limits.
- Require that market prices reflect the real costs of production by integrating social, environmental, and ecosystem values into pricing.

- Ensure greater materials efficiency and circularity by 2025.
- Actively support efforts to restore degraded lands and water through methods such as open ocean plastics recovery and Ecosystem Restoration Camps.
- Recognize that the degree of social change needed to make a successful transformation to a sustainable future will extend throughout society, requiring fundamental shifts in behavior and rethinking of national and community support and care systems.

For more than a quarter century, world leaders, scientists, and expert advisers have been meeting to try to do something about climate change and the other tragedies of the commons, first chronicled in the Club of Rome's "Limits to Growth" study in 1972. These conferences and meetings achieved international treaties, including the Biodiversity, Deforestation, and Desertification Conventions and the Paris Agreement on climate change. But what has not changed is the trajectory of the crisis or the common understanding of interconnection and reciprocity, something natural scientists have understood since the era of Humboldt, Bolívar, Darwin, Haeckel, Marsh, and Muir in the nineteenth century. Rather than preserving biodiversity and forests, we are now well into a Sixth Great Extinction event, losing forest cover and desertifying faster than ever before. Rather than moving toward carbon neutrality, greenhouse gases are still growing, and the rate of yearly emissions even accelerated from 2016 to 2018.

Our problem seems to be the inertia of bad decisions made in the past. But humans can and do change their patterns of living, and that can most readily be seen in the world of fashion.

A few years ago I was teaching a permaculture design course in Estonia when one of my students, fashion designer Reet Aus, asked if fashion had a role in permaculture. Most mass-production manufacturers send about 18 percent of pre-consumer textiles as scrap to a landfill or incinerator. Aus's PhD dissertation was "Trash to Trend: Using Upcycling in Fashion Design," which opened up new possibilities within the fashion industry. Since 2002, Aus has been upcycling—turning unwanted materials into new, mass-produced garments. Her Bangladeshi partners sweep up floor cuttings from Tommy Hilfiger, Bershka, Calvin Klein, and Zara to add into her latest designs.

Her collection, including a treasured shirt of mine, is entirely from postproduction leftovers. She keeps proving that clever design can salvage mountains of wasted textiles and the labor and natural resources spent to produce them, usually inside the same factory. Each garment in her line will save on average 75 percent in water and 88 percent in energy. She also improves the working conditions of the shops she helps in Bangladesh.

"In my opinion, we should keep oil-based fabrics in the loop as long as possible," she told me recently. "Clothing in this area is not the biggest problem, but we can just stop making fabric from oil. We have a lot of good alternatives, from algae to cellulose."

Aus has tapped into an element of human nature that has led to our present predicament but could also point to the way out. It isn't science or technology that confounds us from rejoining Earth's ecology; it's social behavior.

As can be seen in zebras or wildebeest crossing a river full of crocodiles, herding is a rational defense strategy. Bunching herds protect their majority from predators, although a few will be lost to the needs of the river dwellers. Millions of years ago, our ape ancestors adopted herd strategy over lone individualism, and it has served us well. Our fads and fashions are not optional—they are hardwired to our genetic code. When we choose to wear a necktie and blazer, or a pantsuit with jewelry and heels, we are signaling membership in a particular band. The cars we drive, the places we live, the foods we eat—all are signals of belonging to one specific tribe.

Tribal instincts toward personal sacrifice are ennobling, unifying, and heroic. In his book *Tribe: On Homecoming and Belonging,* Sebastian Junger writes:

> The human conscience evolved in the middle to late Pleistocene as the result of the hunting of large game. This required cooperative, band level sharing of meat. Because tribal foragers are highly mobile and can quickly shift between different communities, authority is almost impossible to impose on the unwilling. And even without that option, males who try to take control of the group or the food supply are often countered by coalitions of other males.

> This is clearly an ancient and adaptive behavior that tends to keep groups together and equitably cared for.

Fashion is how we signal not merely tribal allegiance but the values we share. When we choose to go plastic-free, whether in our clothing or the packaging and transportation of the things we exchange, we signal membership in the next order of humans on Earth: *Homo regenesis*.

Refuse

The sudden emergence of plastic in the twentieth century caught evolutionary biology by surprise. While human brains may be well designed for collaboration, our senses are not keyed to respond easily to large, slow-moving threats. We have what psychologists call normalcy bias.

According to a 2014 article in *The Guardian*:

> "Our brain is essentially a get-out-of-the-way machine," Daniel Gilbert, a professor of psychology at Harvard best known for his research into happiness, told audiences at Harvard Thinks Big 2010. "That's why we can duck a baseball in milliseconds."
>
> While we have come to dominate the planet because of such traits, he said, threats that develop over decades rather than seconds circumvent the brain's alarm system. "Many environmentalists say climate change is happening too fast. No, it's happening too slowly. It's not happening nearly quickly enough to get our attention."

Humans are saddled with other shortcomings, too. "Loss aversion" means we're more afraid of losing what we want in the short term than of surmounting obstacles in the distance. Our built-in "optimism bias" irrationally projects sunny days ahead in spite of evidence to the contrary. To compound all that, we tend to seek information not only for the sake of gaining knowledge but also for the purpose of reinforcing already-established viewpoints.

Two types of cognitive bias—confirmation and normalcy—beset us. We seek out and assign more weight to evidence that confirms our views of the world—views we mostly formed as children as we "aped" parents, teachers, inspiring leaders, and celebrities. Our fondness toward normalcy lets us box out things that make us feel uncomfortable and allows us to focus on ways to blend into the crowd. If the crowd thinks peak oil, climate change, the moon landing, Kennedy's assassination, or the inside job at the World Trade Center are just weird conspiracy theories by crazies at the fringe of our society, we ape the crowd. That's just our social software.

Considering that human minds are capable of great feats of irrationality, is there really much hope we will respond quickly enough to the emerging but slow-moving threats of plastics, environmental radioactivity, petrocollapse, or climate change?

> Paranoia? Of course not. It's alternative scholarship. What's wrong with teaching alternative theories in our schools? What are liberals so afraid of? . . . Why this dictatorial approach to learning anyway? What gives teachers the right to say what things are? Who's to say that flat-earthers are wrong? Or that the Church was wrong to silence Galileo, with his absurd theory (actually written by his proctologist) that the earth moves around the sun. Citing "evidence" is so snobbish and élitist. I think we all know what lawyers can do with evidence.
>
> **ERIC IDLE,** "WHO WROTE SHAKESPEARE?"

We are accustomed to most threats being reversible or avoidable. We are accustomed to getting ample warning so we have time to consider and need only act once a problem becomes big enough or close enough to be really, really scary. We also like simple answers to hard questions. Following the example of René Descartes, we prefer to take complex phenomena and break them into categories. Fuzzy continua get broken into inches and pounds. We might be able to work out solutions to some problems that way, but we are more likely to miss the bigger picture that requires us to grasp relationships.

Our linear cognition evolved before we came down from the trees, when you could plot a course three branches ahead, like Tarzan, but if you projected your mental map to a fourth branch, there was a good chance you might miss the nearest one while you were so deep in thought.

Nonlinearity and quantum phenomena puzzle us. How is it that prey can sense they are being observed even when there is no sight, sound, or smell to reveal their predator? Our pattern recognition only extends to "as before, so thereafter," or even "after this, therefore because of this" (i.e., "stocks were down today on growing discomfort from trade sanctions"). We can't ken that when something jingles over here, something unrelated jangles over there.

So it is that when ice in the Arctic describes a superlinear melt curve or wildfires level whole neighborhoods in California, we are so dumbfounded we are more than willing to accept that it's just the weather.

The European Commission, in a 2011 "Plastic Waste in the Environment" report, described these trends as likely to continue:

- There will be an upward trend in demand for plastics.
- Plastic waste will increase.
- Levels of recycling, primarily mechanical, will increase.
- Levels of energy recovery will improve but in a more limited way than recycling levels.
- The level of exports of waste, in particular, plastic waste for recycling and recovery, looks set to increase as overall recycling levels and volumes increase.
- The production of plastics will also tend to be dominated by the Asian market, particularly China.
- The production of bioplastics, while remaining a relatively low proportion of total plastic use, will increase rapidly, waste-to-energy processing (incineration) is set to grow, and landfilling may decline.

They were correct about many of those points but wrong on two important counts. Landfilling remains a strong option and may increase, and when China ended its imports of recyclables, exports from Europe and elsewhere began not just to slow down but to stop. There are other countries, particularly in Asia, still willing to accept recyclable plastics, but most of those have already announced plans to follow China's lead.

Until we grew to be seven billion, going on eight, the world was big enough that there was somewhere we could think of as *away*. Most of the world was an ocean. Cities could barge their trash out to sea and just dump it. Now even the ocean is too small. It is finite, while the capacity of *Homo colossus* to consume and pollute is exponential. Sooner or later, and later is now, those two rates have to meet.

What can you do? Do without. Reject plastic in your life. Camille Scheidt, author of *Unraveling Threads: How to Have a Sustainable*

Wardrobe in The Age of Plastic Fabric, says, "We must learn to appreciate better clothes and less of them. Just like you would for a meal, or most other products you buy, you need to start asking, 'what is this made out of, where did this come from, and who made this?' . . . Often, we'll use the excuses of 'well, this is recyclable, isn't it?' or 'I'll just donate it,' to justify buying more . . . those excuses don't cut it, and neither recycling nor donating clothing is as sustainable as they seem."

You can start rejecting plastic in your life by simply refusing to be served a single-use plastic straw. You can buy only wooden toys and home furnishings. Bag groceries in paper, if not reusable cloth. Buy biodegradables. If there is to be a future, this is where it begins.

Michelle Cassar told *My Plastic Free Life* blogger Beth Terry that she had personally refused more than ten thousand items in the previous three years. Cassar wrote:

> My boyfriend and I live in a rural town on the west coast of Portugal, just minutes away from the Atlantic. All totals are approximate. But I have under- rather than over estimated.
>
> Plastic bags *(including produce bags, bin bags, bread bags, bags to put the bags in bags)* @ 25 per week = 4,000
>
> Bottled water in 5L bottles *(for six months of the year while living in a camper van)* @ 4 a week = 288 bottles, 288 bottle tops, and the plastic that wraps the bottles together
>
> Small bottled water out and about @ 2 a week = 312 bottles, 312 tops
>
> Shampoo, shower gel, conditioner @ 1 of each every 3 months = 36
>
> Deodorant @ 1 per month = 36
>
> Mouthwash @ 4 a year = 12
>
> Cotton buds/Q-tips @ 2 boxes of 200 per year = 1,200
>
> Plastic cups at events and in bars, approximately 10 times a year, 4 drinks a time = 120 *(They won't refill a plastic cup because of "hygiene," so for every person they have a new cup with every drink.)*
>
> Plates, crockery at events. Countless! At one event alone, I refused about 10 items, and we have loads of outdoor events here.
>
> Straws approximately @ 3 a week = 500
>
> Boxed wine *(i.e., plastic bagged wine with plastic tap)* @ 2 a month = 72
>
> Fruit punnets @ 4 a week = 624
>
> Pasta bags @ 2 a month = 72 *(The boxes still have a small plastic window, but it's far less.)*

Crisp packets @ 3 per week = 468 *(I still eat crisps; the Portuguese ones are so good. Now it's a rare treat, maybe once every 2 months, rather than 3 packs a week like before, which is probably better for me too!)*

Peanut butter @ 1 a month = 36

Tomato sauce @ 2 a month = 72

Butter tub @ 1 a month = 36 *(Though I'm sure I eat more butter, I would rather not admit it to myself!)*

Mayonnaise, brown sauce @ 2 a month = 72

Cheese @ 1 a week = 156 packets

Mini butter sachets / tomato sauce / mayonnaise when I eat out @ 4 a month = 624 *(Arghhh. I love butter, so this is always a hard one.)*

Bottled water while eating out @ 2 a month = 72 bottles, 72 tops

Tampons @ 3 a day, 7 days of a month = 756 *(I now use a mooncup.)*

Cling film, 2 boxes a year @ 5 meters = 30 meters

Clothes. All clothes are delivered to the shops individually bagged, and a lot of them are made from plastic fibers. I've not bought any new clothes in over three years. *(A big change from my student cheap high-street days!)* How many items in three years? A lot! Approx 4 items per month = 144 *(Now I go to charity shops, boot sales, and have lovely friends' hand-me-downs.)*

Takeaway cartons @ 2 a week for 4 months = 8

Total: Over 10,389 items (!) + crockery at events.

In *Zero Waste: 50 Tips for a Plastic-Free Life,* author Caroline Piech offers ways for everyone to change their habits immediately. Here are some of her recommendations that haven't already been mentioned:

- Stay away from viscose, nylon, spandex, and polyester.
- Learn how to handle a needle and thread.
- Cosmetic products should best be stored in glass, metal, wood, cardboard, or paper containers.
- Choose toilet paper that is not wrapped in plastic.
- Use washable makeup pads.
- Farmers markets and market days are a great way to buy fresh, local produce without plastic packaging and stickers.
- Buy fresh eggs in cardboard boxes, not polystyrene.
- Stop using disposable coffee pods or capsules.

- Use wooden or paper matches or refillable metal lighters instead of plastic disposable lighters.
- Many popular chewing gum brands are overloaded with plastic, synthetic rubber, and artificial sweeteners. Avoid chewing gum!
- Use glass, tin, or steel containers for food leftovers instead of Tupperware.
- Acquire absolutely necessary plastic articles secondhand.
- Use natural wipes and scrubbers instead of plastic dusters and synthetic sponges.
- Replace cutting boards, cutlery, dishes, strainers, mixing bowls, etc. with natural materials, such as wood, metal, bamboo, or compressed cotton.
- Wash your laundry with homemade laundry soap.
- Never flush cat litter down the toilet.

It is easier for our limited, linear brains to make the right choice if we get some guidance that is timely and easy to understand. Ekoplaza, a Dutch supermarket chain, has created the first plastic-free supermarket aisle with seven hundred items without plastic packaging. They are packed in compostable biomaterials, cardboard, glass, and metal. Another Ekoplaza innovation is a new label that reads, in bright lettering, "Plastic Free." Why should we not have this everywhere?

The Berlin food company Foodpanda, part of pizza.de, offers its customers in the Asia-Pacific region the choice to decline disposable cutlery. In Singapore, about 10 percent of customers declined, which saved one million pieces of plastic cutlery within the first year. But why offer the plastics choice? Why not just provide starch-based cutlery and offer the choice to decline that? These companies are getting pats on the back for taking half measures. Not even half. Ninety percent of Foodpanda customers are still receiving throwaway plastic, and that is just in markets where a choice is offered.

When providers are still behaving irresponsibly, the burden falls upon purchasers to do the right thing, do their homework, and turn away from patronizing bad actors.

CHAPTER 7

Learning to Live with Plastic

> In a new plastics economy, plastic never becomes waste or pollution. Three actions are required to achieve this vision and create a circular economy for plastic. *Eliminate* all problematic and unnecessary plastic items. *Innovate* to ensure that the plastics we do need are reusable, recyclable, or compostable. *Circulate* all the plastic items we use to keep them in the economy and out of the environment.
>
> **ELLEN MACARTHUR FOUNDATION**

The Ellen MacArthur Foundation's New Plastics Economy (NPE) project sets a vision of a circular economy for plastic, one in which plastic never becomes waste.

> For plastic packaging, specifically, we recognize a circular economy defined by six characteristics:
>
> 1. Elimination of problematic or unnecessary plastic packaging through redesign, innovation, and new delivery models is a priority.
> 2. Reuse models are applied where relevant, reducing the need for single-use packaging.
> 3. All plastic packaging is 100 percent reusable, recyclable, or compostable.
> 4. All plastic packaging is reused, recycled, or composted in practice.
> 5. The use of plastic is fully decoupled from the consumption of finite resources.
> 6. All plastic packaging is free of hazardous chemicals, and the health, safety, and rights of all people involved are respected.

Unfortunately, as I dug more deeply into the NPE manifesto, I began to see it as being unrealistically demanding in many ways. It assumes we can invent our way out of this problem with new generations of bioplastics, and although that could be true for some products, it is not universal, nor will it be quick or inexpensive.

The Linear Model, from "Circular Economy 2.0" by Alexandre Lemille. Drawings by Rachael Acker.

I can agree that the elimination of problematic or unnecessary plastic through redesign, innovation, and new delivery models should be a priority. Single-use plastics are the low-hanging fruit, and many cities and countries are now moving to ban those. But when the New Plastics Economy calls for all plastic packaging to be 100 percent reusable, recyclable, or compostable, there is going to be a problem, because in some cases those qualities will make the packaging unsuited to its purpose. Replacing plastic packaging is not a blanket solution; it is one for specific, targeted applications.

Let's review what we have learned so far:

Of the different varieties of polymers in the marketplace today, some are made from agricultural wastes or renewable sources, such as algae, but most—90 percent—are not.

Not all bioplastics are also biodegradable. Most are not, and many that are gum up existing recycling plants when they get commingled with PET or other recyclable plastics. Still, we should choose biodegradables whenever the opportunity presents itself to help that market grow. Those of us who can compost these ourselves, should.

Of the remaining legacy plastics, these are separately recyclable:

PET (polyethylene terephthalate) is clear and tough and a good barrier to moisture. When woven into fabric, it's known as polyester. PET is commonly used for beverage bottles, condiment bottles and jars, frozen food trays, clamshell containers, mouthwash and other toiletry bottles, clothing, and carpeting.

HDPE (high-density polyethylene) is a rigid plastic that resists corrosion. It's used for milk and juice jugs, laundry detergent and bleach bottles, shampoo and other toiletry bottles, vitamin bottles, outdoor furniture, and playground equipment.

LDPE (low-density polyethylene) is a transparent, flexible plastic used extensively in wraps, bags, and other films. It's used to wrap packaged products, such as cases of water bottles and diapers and in cling wrap for food protection, squeezable condiment bottles, and mail-order packaging.

PP (polypropylene) is a durable plastic that can stand up to a range of temperatures and substances and is commonly found in yogurt and margarine containers and lids, deli containers, beverage bottle caps, clamshells, medicine jars, condiment bottles, convenience store drink cups, and reusable food storage containers.

PS (polystyrene) is a versatile plastic that can be made rigid and clear or into an opaque foam. It's in clamshell containers, plates, cups, lids, cutlery, meat and poultry trays, protective packaging, and building insulation.

Nylon polyamides are easy to mold and make into fibers and weave into fabrics. If they can be separated, they can be recycled.

These other types of plastic might technically be able to be downcycled, but because they have toxic compounds or other practical barriers to recycling, until we discover worms, fungi, or bacteria that can safely eat them, they need to be landfilled:

PE (polyethylene) can be easily melted if time and energy are spent to collect, transport, clean, segregate, and burn it. Unfortunately, two by-products of the melting process are CO_2 and water, both major greenhouse gases. One exception is polyester fiber, which can be reclaimed and recycled, even from blends, by a nontoxic chemical process.

PVC (polyvinyl chloride) offers clear, high-impact strength and resistance to corrosion. It is in packaging such as clamshells as well as in rigid film, pipes, siding, flooring, window frames, insulation for electrical wire, hotel key cards, and credit cards. It is toxic both in its manufacture and in the additives it contains to give it plasticity.

PC (polycarbonate) is light, strong, and temperature tolerant. If made from plant-derived carbon, it is a drawdown technology. It is in optical storage media, eyeglasses, shatterproof windows, and airplanes. The process of synthesizing it is dangerous, however, because it is produced by the reaction of phosgene (a chemical weapon) and bisphenol A (an endocrine disruptor).

Landfills and Incineration

Landfills get a bad rap, and in many cases that may be deserved if they are poorly engineered, poorly managed, or leach or outgas poisons or greenhouse gases. But they still are going to have a role as we find ways to deal with our plastic mistakes of the past. Without them, carbon that had been stored deeper underground for millions of years as fossil carbon will, over the course of centuries, be wafting up to the atmosphere to add to global warming.

This is not an insoluble problem if handled responsibly. Biochar is a food-grade charcoal produced from agricultural by-products by pyrolysis (heating in the absence of oxygen). Added to lacto-fermentation vats, it speeds anaerobic digestion and produces a "compost tea" of effective microorganisms (EM). When that biochar-EM mixture is added to compost, along with any needed minerals, it yields a potent biofertilizer that both rejuvenates soils and sequesters carbon. Adding biochar-EM also significantly reduces or eliminates the formation of greenhouse gases CO_2, N_2O, and SO_2 during composting and, when fed to cattle or poultry or added to biodigesters, significantly reduces methane, another potent greenhouse gas.

Biodegradable plastics should be composted in a manner that is less likely to outgas methane, such as with the two-stage process of biochar-assisted wet lacto-fermentation in an anaerobic environment followed by biochar-assisted dry thermophilic composting in an aerobic environment. This bacteria/fungi two-step process not only curtails methane but also can reduce other greenhouse gas emissions, such as nitrous oxide and sulfur dioxide.

Similar issues arise when we consider incineration, including the very promising technique of carbon drawdown by electric cogeneration from waste biomass-to-biochar systems (called BEBCS in the negative-emissions world) followed by soil or hardscape sequestration. We don't want to overlook the potential for extracting the embodied energy in polymer molecules, rather than just letting microbes digest them. Plastics burn hot in waste incinerators, equal on a weight basis to the heat output of diesel. But then the fossil carbon (unless scrubbed by expensive filters) is released into the air immediately as greenhouse gases. Michael Tolinski, a former editor for *Plastics Engineering*, reminds us that "conceptually, burned biopolymers provide a kind of 'solar energy' from carbon captured via photosynthesis, as in the burning of wood."

"But in reality," Tolinski continues, "bioplastics' energy from incineration would mainly compensate for the extra fossil-carbon-based energy used during biopolymer synthesis. Overall, in incineration, bioplastics would still seem to be favored over fossil fuel plastics, though future life-cycle studies on this issue would be helpful."

Right now, PLA and PHB are in a gray area between the recycling plant, landfill, or incinerator. They can be plant-based bioplastics. They can even be synthesized from bacteria and algae. They show a lot of promise for the plastics of the future. However, because they can gum up today's recycling lines and because they can't be mechanically separated from PET bags that look and weigh the same, they have been banned in some places until recycling issues are resolved.

Our Choices

I led this chapter with a quote from the Ellen MacArthur Foundation calling for the elimination of "problem" plastics and replacement of the ones we keep with biodegradable or recyclable versions. The worldwide volume demand for plastics packaging will not decrease anytime soon; rather, the opposite is happening. However, the industry is responding to calls like MacArthur's to develop better, cheaper, bio-based plastic materials for packaging.

Plastics used in construction are nearly the exact opposite of plastics used in packaging in that they are meant for long-term use rather than relatively brief, disposable uses. As packaging, they are expected to be small, lightweight, portable, and possibly transparent. As wiring, plumbing, decking, or floor tile, they are expected to be durable,

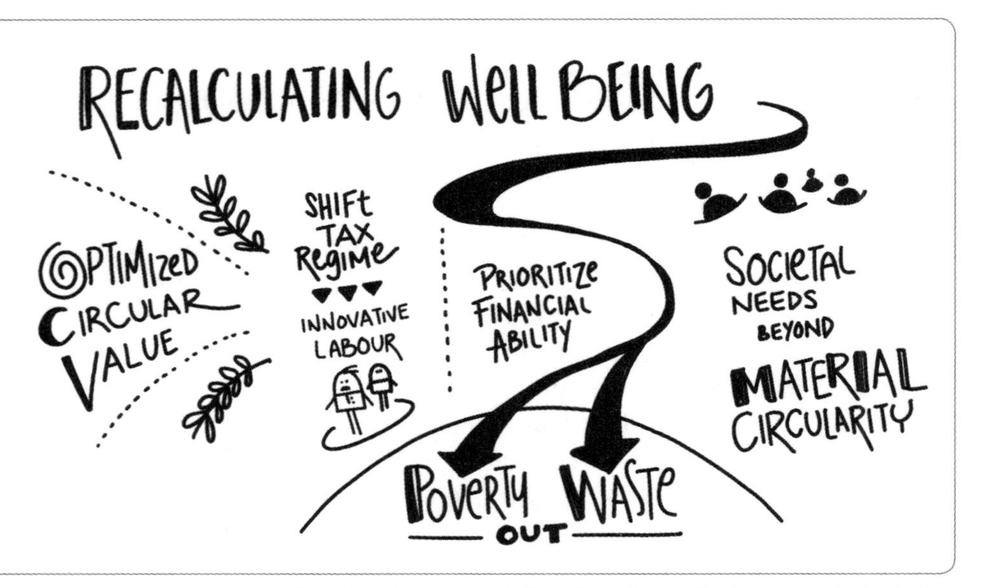

Reinvisioning, from "Circular Economy 2.0" by Alexandre Lemille. Drawings by Rachael Acker.

fireproof, and insulating or ventilating, and to possibly mimic the look of wood or stone.

Some uses for plastic are practical necessities, even when they are not recyclable, bio-sourced, or biodegradable. New plastic composites are allowing cars, planes, and trains to go farther using less fuel. Solar cells are being fabricated as thin polymer films that can be painted on rooftops or layered onto clear glass windows. Landfill liners, sewer pipes, flame-resistant electrical boxes and cables, and many other common items have been widely adopted for safety reasons. You can complain about the Ziplock bag, but the truth is that it saves a lot of food from spoiling, and indirectly it lowers the cost of food. In city hospitals and rural clinics, medical plastics are the only materials sterile enough to rely upon precisely because microbes don't like them. And, what heart patients would want to have a biodegradable pacemaker in their chest?

These, then, are our choices. We can choose better plastics and reject worse ones. Bio-based plastics should come from biological waste products rather than go into competition with food crops and forests. We can recycle what can be recycled and then properly dispose of that which cannot.

We need to clear up the morass of labeling to make it less confusing for people who want to recycle. As Tolinski says, "Little can be gained in the long run by having consumers think that any plastic product imprinted with the 'chasing arrows' or equivalent recycling symbol can be or is being recycled."

Plastics are not just going to disappear overnight. They are too deeply embedded in our global civilization.

As ecological activist Julia Butterfly Hill says, "The question is not 'can you make a difference?' You already do make a difference. It's just a matter of what kind of difference you want to make during your life on this planet."

A more well-ribbed formula for creating the circular economy might be found in six "safe" and "just" principles drawn from Ellen

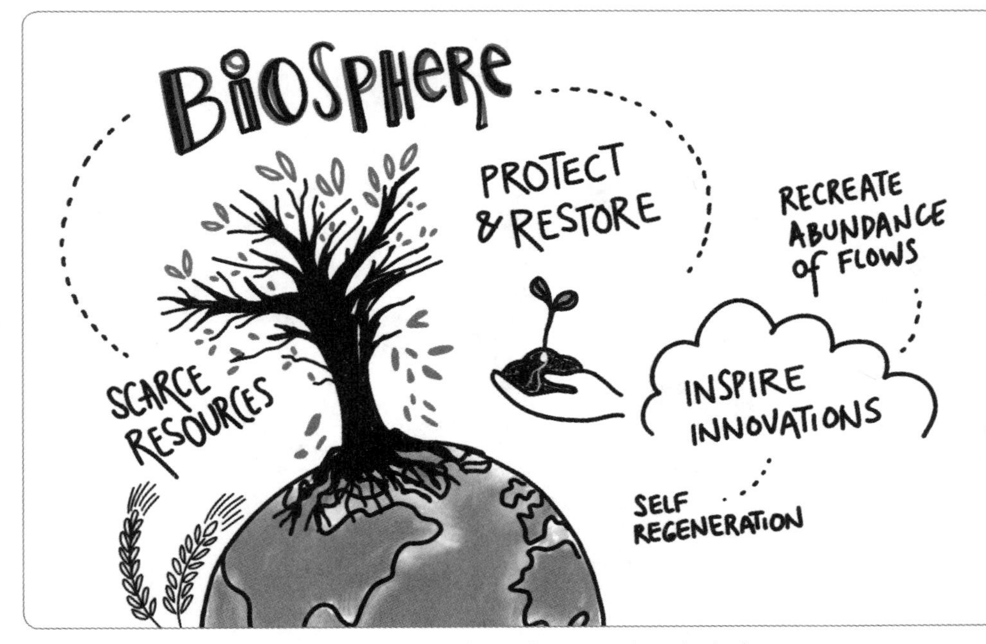

Finding Context, from "Circular Economy 2.0" by Alexandre Lemille. Drawings by Rachael Acker.

MacArthur, then adapted by Alexandre Lemille and Tom Harper in a 2018 article for *The Beam*, a triannual magazine published in Berlin:

- Safe Principle #1: Preserve and enhance Natural Capital by controlling finite stocks and balancing renewable resource flows.
- Safe Principle #2: Optimize resource yields by circulating products, components, and materials at the highest utility at all times in both technical and biological cycles.
- Safe Principle #3: Foster system effectiveness by revealing and designing out negative externalities.
- Just Principle #4: Equity makes business sense if services are designed to address the needs of all.
- Just Principle #5: Developing people's means of exchange is a priority for accessing more with less in a service-based economy.
- Just Principle #6: In a systemic regenerative model, all abundantly available renewable energies should be considered, including labor.

Lemille and Harper propose a "Circular Economy 2.0"—eradicating both waste *and* poverty. Their idea is a Circular Humansphere using Natural Capital, Human Capital, and Remanufactured Capital, as well as only abundant or endless flows of available energies. They are especially fond of economic system reform based upon distributed ledgers,

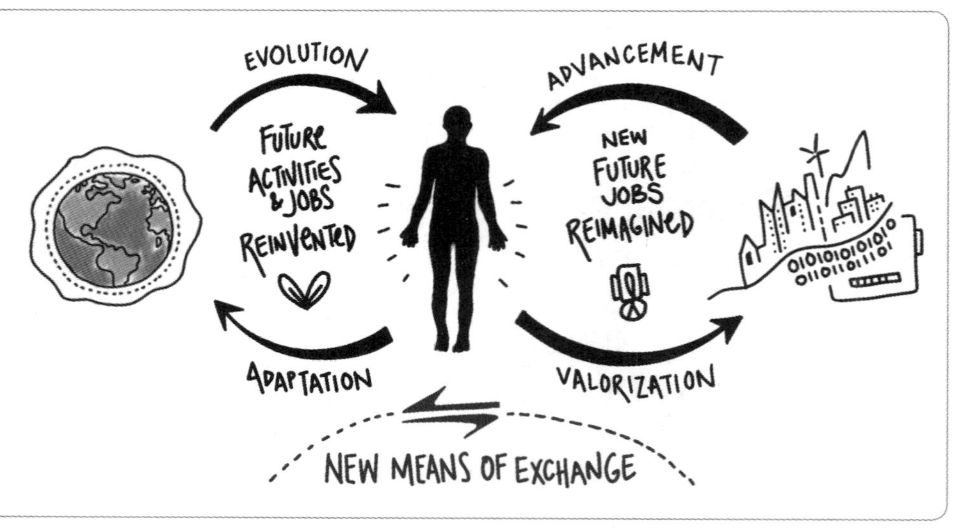

Redesigning, from "Circular Economy 2.0" by Alexandre Lemille. Drawings by Rachael Acker.

such as Blockchain (although one should take note of how energy-dependant and technologically brittle the present Blockchain system is). They write:

> Up until now, price has determined our values in the marketplace, often at the cost of social and planetary boundaries. Imagine a system where the value of the transaction is determined by the level of impact a good or a service has on the planet and its people. Imagine being able to transact intangible values, such as love, well-being, harmony. Imagine being able to digitize the impact of our transactions, recording transparently and immutably those transactions which align with the key principles of a Circular Economy and those that don't.

Whether we will ever see an "internet of value," each of us can make consumer choices that value the earth and its people as if they mattered more than money and profits. The problem of plastic turns out to be less about polymeric chemistry than the suicidal ethos of a throwaway culture. We must individually and collectively choose to eradicate the exponentially extractive and destructive patterns we have grown so fond of only in the past three or four generations. Perhaps our best way to do that is by unlocking the collective bounty of irrepressible creativity within the human spirit.

CHAPTER 8

The (Only) Way Out

It can be difficult to tell the difference between things falling apart and things falling together.

JANE LEWIS

On November 28, 2018, the Ethical Corporation, a manufacturing industry think tank, convened an online panel discussion, "Innovation in Plastics: Reduce, Reuse, and Recycle," that was downloaded and viewed by thousands of interested participants around the world. Panelists included raw plastics refiner Dow, mid-level formulator Amcor, and end-product designer Walgreens. Enthusiastically, each spokesperson, bearing titles like "Vice President for Sustainability" or "Manager of Innovation," endorsed optimistic goals, such as all packaging being recyclable by 2025, a measure put forward by the New Plastics Economy in 2015 that hundreds of manufacturers have now endorsed. There was just one problem: there is as yet no practical roadmap for how such ambitious goals might be achieved.

These thoughtful people at the top of the plastics industry said nothing that offered a real way out. Instead, they talked about hope for a new generation of biodegradables and ambitious recycling targets that failed to take into account the recent decisions by Asian countries to stop allowing themselves to be a dumping ground.

There is a solution—don't get me wrong—but it won't be found in mere promises to do better.

I am reminded of the famous Walt Kelly cartoon from Earth Day in 1971 in which the little possum, Pogo, is walking through his swamp, now filled with plastic rubbish, with his friend Porkypine. "Ah, Pogo," says Porkypine. "The beauty of the forest primeval gets me in the heart."

"It gets me in the feet, Porkypine," replies Pogo.

"It *is* hard walking on this stuff," says Porkypine.

"Yep, son," says Pogo, "we have met the enemy and he is *us*."

At the beginning of this book, I told you about Albert Bartlett's famous lecture on the exponential function. Bartlett might say the "us" Pogo was referring to is the species that is smart enough to land a mixed-media polymer rover on Mars but unable to grasp the arithmetic of the exponential function and how that applies to the nonrenewable resources of Earth. We find ourselves as a race that started with small bands of upright hominids moving out of the trees and then out of Africa, but whose DNA is hardwired into defending a smaller, "normal" past that no longer exists.

© Okefenokee Glee & Perloo, Inc. Used by permission. Contact permissions@pogocomics.com.

Prisoners of our genetic heritage, we are followers of fashion, fiercely loyal to our psycho-demographic (which may not be familial but exists simply because we wear the same football jerseys or practice brand loyalty in choice of automobiles, foot-

Fitting In, from "Circular Economy 2.0" by Alexandre Lemille. Drawings by Rachael Acker.

wear, or political party). We are happiest hearing the opinions of those who will confirm our beliefs or denigrate beliefs that differ. We are eager to deal with problems that will have consequences in the next ten minutes or ten hours but disinterested in consequences that will affect us only some years and decades later.

In his seminal treatise on this subject, *The Wizard and the Prophet,* Charles C. Mann divides our tribal affinities between those who believe human ingenuity and technological progress will eventually solve all the predicaments besetting us (the Wizards) and those who see the source of all our problems as a separation from the natural world (the Prophets). While Mann has nice things to say about each of these viewpoints, I must disclose that I lean strongly toward the Prophet camp.

Nature provides the way, as it always has. It has the advantage of two billion years of trial and error going into the results we see all around us.

Solutions at the scale now required won't come through Cartesian dissection—breaking things into component parts and trying to rearrange them better. Solutions will come, if at all, only by approaching the problem as one of wholes, as nature does. Our problems are crosscutting, nonlinear, synergistic, and compounding. So must be our solutions.

One of the issues with plastics that must be addressed is our dependence on feedstocks that may or may not be capable of being sustainably produced. Fossil fuels are teetering on the brink of industry-wide collapse, either because of natural depletion (in less than two centuries we used up most of the low-hanging fruit, and in this century we are being forced to spend more each year to pump up and refine the dregs, a costly and time-limited proposition), debt (we have compensated the decline in low-cost energy by paying subsidies to producers at a bankrupting scale—some $6 trillion in 2018 and escalating), and cascading environmental consequences, first and foremost rapid-onset climate change. California recently announced that it is joining a handful of other large economies around the world with plans to begin curtailing fossil fuel *production* by 2025. The end of the age of oil is now upon us.

That leaves us with bioplastics, but therein lies another conundrum. While some polymers might be derived from crop residues, bioplastic formulations that compete for soil, water, and favorable growing conditions with food will have obstacles thrown in their path, as we have seen in the competition for agriculture coming from the corn-based ethanol experience. This will only be compounded by climate change and continued population growth.

Even if waste-sourced bioplastics and recycled resins can replace fossil plastics, there is still a toxic legacy to be confronted and lingering questions about whether poisonous products released in the past can be degraded quickly and efficiently enough to prevent catastrophic damage in the future.

The New Plastics Economy says, "No plastic should end up in the environment. Landfill, incineration, and waste to energy are not part of the circular economy target state." As we have seen, however, landfill, incineration, and waste-to-energy processes may all have a role going forward. This is especially true when we speak about the disposal of nondegradable, toxic plastics and the circular economy of waste biomass to energy plus biochar (via pyrolysis), a drawdown strategy that can work at the local community scale and achieve many sustainable development goals at the same time.

Where we can agree is that, over time, inputs from nonrenewable sources must be switched to renewable sources. The production and recycling of plastic should be powered entirely by renewable energy, plastic packaging should be free of hazardous chemicals, and the health, safety, and rights of all people—and nature—should be respected.

We are witnessing the birth of that future. We are over the peak on the limits to growth and into the next portion of this great cycle of civilization. There is jockeying for position for vanishing scraps on the assumption that there is still some way to regain the peak we only just experienced. That is our normalcy bias. Don't worry; it's in our DNA. But the prosperous way down is to downsize the population along with expectations and mechanistic productivity. We must grow from reckless juveniles into responsible adults.

In my view, three drivers are needed, in combination, to bring about the switch to the next part of the cycle—a truly regenerative human civilization. These three are regulators, money, and conscience.

Regulators

I say *regulators* because having a regulatory scheme and laws on the books is not always enough. You need manpower in the field with the motivation and authority to enforce the rules.

Edmund Burke, writing in *Reflections on the Revolution in France* (1790), said, "The temporary possessors and life-renters in it, unmindful of what they have received from their ancestors, or of what is due to their posterity, act as if they were the entire masters; that they think it amongst their rights to cut off the entail, or commit waste on the inheritance, by destroying at their pleasure the whole original fabric of their society; hazarding to leave to those who come after them, a ruin instead of an habitation—and teaching these successors as little to respect their contrivances, as they had themselves respected the institutions of their forefathers."

Another way to put this would be to say that regulation begins at home. Society is a constant collaboration. Where it works well, as Confucius first observed in the fifth century CE, people prosper in safety and happiness. Where it breaks down, lives are quickly reduced to squalor and misery.

Confucius's teachings of *jen*, *yi*, and *li* are illustrative. *Jen* has been translated as love, benevolence, humanity, human-heartedness, virtue, moral rectitude, and humaneness; it also signifies the ideal relationship between people. *Yi* generally means righteousness, appropriateness, obligation, and justice, and is "the principle of setting things right and proper." *Li* indicates ceremony, rites, decorum, courtesy, etiquette, rules of propriety, and in the full sense connotes the sociopolitical order. Confucius argued that rulers can no more rule an unruly people than drivers can choose whichever side of the road best suits them and not expect to have collisions. Governance comes from the bottom up, not the top down.

Chile made history on August 3, 2018, when it became the first country in Latin America to ban the commercial use of plastic bags. Large businesses were given six months to phase out the use of plastic bags, while smaller ones were allowed two years. That was a top-down decision—a legislative decree. Companies will be limited to handing out a maximum of two plastic bags per transaction during the respective grace periods. In a country where the minimum wage is just $800 per year, those flouting the ban will be subject to a $370 fine.

To balance the bottom-up side of the equation, Chile's wise president, Sebastian Pinera, took to the streets and started handing out cloth bags.

"I want to share with you the joy that as of today we're enacting the law," said Pinera at a public ceremony in the center of Santiago. "Without a doubt, we're taking a giant step toward a cleaner Chile." Pinera then stood and handed out cloth bags to anyone who wanted them. Leading by example is how regulators win the hearts of the people.

In September 2018, California became the first US state to implement a partial ban on plastic straws. Dine-in restaurants no longer are allowed to provide customers with straws automatically. Instead, customers who need plastic straws have to request them. Restaurants that violate the ban will receive warnings first, and repeat offenders will be fined up to $300. Other states and countries have enacted or are considering enacting similar and even more extensive bans. None of these laws would have happened if the people of those jurisdictions had not first demanded them of their government officials.

In December 2018, Boston's plastic bag ban began for retail spaces of twenty thousand square feet or larger and will roll out to smaller businesses over the next few years. The Tamil Nadu government in India announced that single-use plastic is banned in the state from January 1, 2019. After two of Australia's biggest supermarket chains announced that they would stop offering single-use plastic bags to their consumers, bag bans were enacted in all but one Australian state. By year's end, there had been an 80 percent drop in plastic bag consumption across the entire country.

The single greatest determinant of success for these new single-use plastic bans is whether regulators have the support of the regulated.

Also in 2018, the Danish toy maker Lego began selling a new eco-friendly line consisting of twenty-five various brick shapes that will resemble nature-inspired products, made out of polyethylene from sugarcane. However, the toy manufacturer admits that this material is not strong enough to make regular Lego pieces, so it is investing one billion kroner (150 million US dollars) to find a way to make all Lego bricks using biodegradable material. Lego also committed to relying on 100 percent renewable energy by 2020 and to a carbon-neutral supply chain. It is promoting recycling by encouraging families to recycle or donate unwanted Lego bricks. This is a case in which a company, although perhaps prodded by its greening customer base, did not wait for government regulation or industry-wide changes but decided to take the initiative for reasons of social conscience.

Before he became president of the People's Republic of China, Xi Jinping was governor of Zhejiang. One year he decided to go on a state visit to the rural villages to assess the needs of the people. What he discovered was a brewing catastrophe. Globalization had been drawing people from the country to the cities for many decades, and government policies encouraged that trend in order to fill the need for a gargantuan factory labor force. China had recognized that this course meant sacrificing agricultural capacity, but like most countries, it was willing to make that trade-off because it figured that it could import food, and a whole lot more, with its newly favorable trade balance.

What Xi Jinping saw nearly broke his heart. Long a champion of "Chinese values" and the "Chinese dream," Xi had hoped to revive Taoist practices of harmony in culture and nature. What he saw in the rural countryside was that all the teenagers, young people, and middle-aged adults had left. There were only the very elderly (the grandparents) and the very young (the grandchildren) being supported by a combination of welfare services and remittances from distant families working in the cities. On land too steep to use machinery, terraces were in disrepair, overgrown with weeds and emergent forest. Buildings were crumbling and stray dogs roamed the streets. Food production had plummeted. The old hand tools were rusted and broken. The forests on the hillsides had been raided by timber companies, and now mudslides wrecked the streams and threatened the villages.

The villagers said to Xi, "Look what we have lost!" They wanted back the forests and wildlife that made this a good place to live. Thus was born Xi's Two Mountain policy.

Back in Shanghai, Xi gave a speech calling for two mountains. The first was the mountain of silver: modern development, including basic services to make people's lives better. The second one Xi called the mountain of gold: return of nature. Pure forests and pure water were the true gold of China.

In 2013, Xi became general secretary of the Communist Party of China, president of the People's Republic of China, and the chairman of the Central Military Commission, the most powerful consolidation of power since before the Tiananmen Square protests of 1989. China's "ecological civilization" concept was first announced in 2007. In 2013, the Central Committee made it part of the Chinese Constitution. In April 2015, China began performing natural resource audits, forcing every official to pay attention to environmental protection while in office or be held to account.

In July 2017, the world's largest importer of recyclable plastics and other waste shocked the world by announcing it would no longer accept other countries' trash and would focus more on curbing its own pollution. Formal protests to the World Trade Organization from Washington, London, and Brussels failed to persuade Xi to change policy, and the new plastic ban took effect on January 1, 2018. The effects are now rippling outward.

The United States, which had been sending millions of tons of scrap plastics to China every year, imposed billions in tariffs on Chinese imports, hoping to force Xi to reverse his plastics ban, among other goals. A Foreign Ministry spokesperson for China said in reply that "it's very hypocritical of the US to say China is breaching its WTO duty. Restricting and banning the imports of solid waste is an important measure China has taken to implement the new development concept, improve environmental quality, and safeguard people's health. We hope that the US can reduce and manage hazardous waste and other waste of its own and take up more duties and obligations."

Commercial and municipal recyclers in the US, Canada, Ireland, Germany, and other exporting countries were left to deal with rapidly growing mountains of plastic. In 2018, the amount of plastic sent to China and Hong Kong fell 92 percent, but scraps being sent to Thailand rose 1,985 percent, Malaysia 273 percent, and Vietnam 46 percent. Then the second shoe fell. Thailand, Malaysia, and Vietnam all announced they would join China's plastic ban. Vietnam immediately stopped issuing new permits for recyclers and will allow existing permits to expire. Malaysia didn't wait and instead revoked the permits of all 114 plastics importers. After a pilot whale washed up dead on a Thai beach in June 2018 with seventeen plastic bags in its stomach, Thailand announced it, too, would join the ban.

Thailand has now established a twenty-year strategy to keep plastic off its beaches and encourage eco-friendly alternatives. Companies such as PepsiCo and Procter & Gamble have pledged $100 million to build out recycling capacity. The reason megacompanies like these are throwing money at Thailand is because they see what the plastic ban will do to their businesses.

Towns and cities across the United States started taking a variety of steps to deal with their mounting backlog. Some suspended recycling schemes, began education campaigns, or refused to accept certain types of plastic waste. In 2018, the US was able to divert two-thirds of the plastic it had been exporting to China to the other Asian countries, but that left some 280 million tons that had been exported before now unaccounted for. It is likely much of that wound up in illegal dumps, lakes, rivers, and the ocean, as consumers unburdened themselves in secret.

"It is an embarrassment that the government of one of the most powerful countries in the world feels it must depend on others to take out our trash," said Greenpeace Oceans Campaign director John Hocevar. "This is a wake-up call for corporations and governments that allow this practice to stop producing packaging and products that no one is willing to dispose of properly. It makes no sense to keep making products that we use once and throw away out of material that lasts forever."

In the wake of the bans across Southeast Asia, thousands of recycling centers in the exporting world are shutting down, unable to make profits from either government contracts or distant buyers. In the wake of those closures, municipal landfills are toughening rules about what can be thrown away to keep from being deluged with diapers that used to go in blue bins. Many municipalities no longer accept plastics numbered 3 to 7, which can include items such as yogurt cups, butter tubs, and vegetable oil bottles.

Ultimately, plastic rubbish backs up to the doors of consumers. What will they do when they can no longer throw that stuff away? After decades of recycling campaigns, many have become "aspirational recyclers" but still can't tell whether a Starbucks cup is lined with plastic or what to do with its obviously plastic lid.

The missing plastic, the failure to anticipate what could happen in China, and now failing to plan for the spreading recycling ban are all symptoms of regulatory failure. Even where laws exist, they are often not enforced. Each fisherman who quietly cuts and loses a torn or tangled net, each cruise liner that dumps discarded single-use plastics overboard, and every city that opens the floodgates from its riverside landfills to let plastics wash out to sea can only do that because of regulatory failure.

Step one should be to hire a new sheriff. Step two would be to inspire better self-regulation. What would a good international regulatory regime look like? It likely would have these characteristics:

1. *Implement a single overarching directorate.*

An increasing number of countries have now imposed a ban on disposable plastics and plastic bags or have established targets for reducing plastic consumption and waste. Why not make these bans and targets uniform and universal? Anyone entering this process can take advantage of the lessons learned by those who went before. Codes can be refined and improved. International meetings can speed the acquisition of knowledge of best practices. Cooperation must be scaled up so that global plastic consumption scales down.

2. *Increase producer responsibility.*

Every two decades world plastic production doubles, and leading plastic manufacturers are planning to increase production by almost one-third over the next five years. In 1974, the average per capita plastic consumption was 2 kg. Today, despite nearly doubling the world's population, it has increased to 43 kg. Producers are taking the world in the wrong direction. The industries responsible for the major plastic wastes must be targeted with specific industry agreements and producer liability arrangements, with requirements for handling, collection, reuse, and disposal.

3. *Increase fees and taxes on polluting plastics.*

Fossil plastic is still less costly to make and buy than renewable bioplastic. Fees must be charged so that recycled, bio-sourced, and biodegradable plastic becomes cheaper than fossil.

4. *Increase waste management.*

The bulk of plastic waste entering the ocean comes from half a dozen Asian countries (the source of 80 percent of marine plastics globally), but rapid population growth and a swelling middle class mean that the consumption of plastic is increasing faster than the ability to handle the waste. It's overwhelmingly clear that at a global level we need to reduce, reuse, or recapture. We need global solutions to supply and packaging approaches and waste collection systems. International aid and technology transfer should be directed to management and recycling infrastructure.

5. Involve people power to clean up the mess.

Estonia lets out schools and businesses for an annual litter pickup day. In Rwanda it happens every month. We should be mapping and surveilling geographical sources and sinks. Cleanup efforts can then be directed to the most likely target zones. We can incentivize a waste-picking culture by placing a value on used plastic that makes it worthwhile to gather. Nature is helping us by collecting plastics in gyres, eddies, and backwaters and on beaches and banks where currents deposit floating flotsam. For example, 38 million pieces of plastic trash litter the shores of Henderson Island, a remote, uninhabited island in the South Pacific. Scientists who surveyed the area believe the island is covered in more plastic trash than anywhere else on the globe. These are places to deploy the best in cleanup technologies and an army of collectors.

6. Increase research.

There is still much we don't know about the plastic problem. We know that more than 70 percent ends up on the seafloor. Over time, it breaks down into tiny particles, but we don't know what happens to this material next. Research on the negative effects and how best to mitigate them must be strengthened. We are only just beginning to discover how fungi and bacteria dissolve even the most recalcitrant organic substances. That work should be supported by an international fund and the results made available widely.

7. Stop the flow of plastic waste into the sea.

About 80 percent of the plastic in the ocean comes from activities and industry on land. The UN has banned this type of ocean pollution, but it lacks the means or methods to enforce its own ban. A central regu-

CAMBIO VERDE

Curitiba, in the southeast of Brazil, has gone from a population of 150,000 in the 1940s to one million in the 1980s and three million today, illegally squatting in ramshackle favelas in houses of cardboard, plywood, and brick, with scant essential services, such as water and power, resulting in corresponding social, economic, and health problems. Trucks couldn't maneuver the narrow streets, steep hills, and deep valleys. People were dumping trash into rivers and fields. Kids were playing in the garbage that was piling up. Trash collection was mostly nonexistent in these barrios until Mayor Jaime Lerner took office in 1972. Lerner changed the civic pride of Curitiba.

Curitiba's Green Exchange Program, Cambio Verde, began in 1989. Today approximately ten thousand Curitibanos collect trash, deposit it at a recycling center, and obtain fresh food, books, football and show tickets, and bus tickets in exchange. For every kilogram of plastic bottles turned in, one kilogram of fresh food is provided.

Throughout the city, two thousand "garis," the men in bright-orange uniforms, sweep streets, sidewalks, and city parks, each cleaning more than a mile per day and together collecting 1,460 tons between sunrise and sunset.

The scavengers and sweepers are only part of Curitiba's proud culture of recycling and progressive environmental policies. Schoolchildren bring plastic to school and get back toys made of recycled plastic. Those children who started with that in the 1970s and '80s are now thirty or forty years older and still have the recycling habit. All public and private schools separate their garbage. Fast-food eateries serve on real plates with real silverware. Once the trash-recycling program was up and running, the city's health department discovered that at least one deadly, mosquito-borne disease was cut by 99 percent.

At the Unidade de Valorizacao de Residuos recycling plant, 230 people sort trash for lunch and minimum wage. Dressed in protective clothing, they toss green glass, clear plastic, milk cartons, yogurt containers, and toothpaste tubes each into its special bin. There are four other government facilities like this one and thirteen private ones elsewhere in the city drawing from twenty-three neighborhood collection centers. Glass becomes glass again, paper becomes paper, metal is turned into screws and nails. Used oils become cleaning products. Handbags are crocheted out of aluminum can flip-tops. An insulation material with good acoustical dampening is made of recycled cement, sand, Styrofoam, and glass bottles. Plastic bottles fitted into each other become chairs. A roof tile made of former toothpaste containers comes with a five-year warranty.

Three months after the program started, with zero budget, 70 percent of Curitiba families were separating trash. Today 70 percent of that trash goes to recycling.

latory authority could do that. In the meantime, governments like that of the state of Parana in Brazil have tried innovative approaches to involve citizens in the cleanup. Parana's governor explained, "If a fisherman catches a fish, it belongs to him. If he catches garbage, we bought it from him. If the day was not good for fishing, the fishermen went for garbage. The more garbage they caught, the cleaner the bays became, and the more fish they would have."

Money

Two hundred and fifty organizations responsible for 20 percent of the plastic packaging produced around the world have now committed to reducing waste and pollution. Five venture capital firms have pledged $200 million toward the initiative, called the New Plastics Economy Global Commitment, a collaboration with the United Nations led by the Ellen MacArthur Foundation. It is composed of a diverse group of members including the city of Austin, clothing company H&M, Unilever, PepsiCo, L'Oreal, Nestlé, and Coca-Cola. Other partners include the World Wide Fund for Nature, the World Economic Forum, the Consumer Goods Forum, and forty academic institutions. Governments that join the commitment must pledge to create policies that help support a circular economy. Corporations must agree to phase out single-use plastic packaging and ensure their packaging can either be reused, recycled, or composted by 2025. Every eighteen months the targets will be reviewed and participating businesses must publish data on progress.

Ultimately, efforts like this are directed at repurposing plastic instead of incinerating it or sending it to landfills. Taken to a global scale, this will require building or improving collection and processing facilities and finding ways to downcycle even the most difficult items into new products.

"While elements of the MacArthur Foundation Global Commitment are moving in the right direction, the problem is that companies are given the flexibility to continue prioritizing recycling over reduction and reuse," said Ahmad Ashov from Greenpeace Indonesia in a press release. "Corporations are not required to set actual targets to reduce the total amount of single-use plastics they are churning out."

Circulate Capital, an investment firm based in New York City that started in 2018, has raised $90 million toward improving plastic waste collection in Southeast Asia and creating markets for collected material. PepsiCo, Coca-Cola, Procter & Gamble, Danone, Unilever, and Dow are all contributors. "There's no silver bullet to stop plastic pollution," says CEO Rob Kaplan. "We're not going to be able to recycle our way out of the problem, and we're not going to be able to reduce our way out of the problem."

Still, he hopes Circulate Capital's investments can serve as one piece of the puzzle. He estimates that more than one billion dollars would be needed to really build out a more efficient waste infrastructure in the region. Circulate Capital hopes to bump commitments to at least $100 million over the next few years.

With major companies taking steps to eliminate or recycle plastic waste, what about what's already in rivers or on beaches? That's where NextWave, a coalition founded by companies such as Dell, Hewlett Packard, and an environmental group called the Lonely Whale, comes in. By employing people living in coastal regions, NextWave collects discarded plastic, primarily nylon 6 and polypropylene, within thirty miles of waterways to prevent it from making its way to the sea. This reclaimed plastic is then shipped to manufacturers who reuse it instead of producing new plastic. Plastic collection sites are chosen based on where cleanup could have the biggest impact and where the plastic could more easily be taken to an existing recycler.

NextWave's partnership with Dell kept a total of three million pounds of plastic from entering the ocean over the past five years. A program with Hewlett Packard worked in Haiti to collect a total of 275 tons of plastic that HP then used to create ink cartridges. According to a press release, HP partnered with a nonprofit called the First Mile Coalition, aimed at improving Haitian labor conditions, to create up to six hundred jobs collecting plastic bottles.

Ikea has announced that it too will partner with NextWave. The Swedish furniture company has committed to phasing out single-use plastics from its stores by 2020 and to designing more sustainably sourced products, including more items made with recycled plastics. Ten companies are now members of NextWave, and they plan to source reclaimed plastics from Indonesia, Chile, the Philippines, Cameroon, and Denmark.

New technology will soon be helping NextWave. In 2018, the Australian city of Kwinana installed a new simple, but unbelievably effective, filtration system to help prevent plastic from reaching the ocean. Kwinana installed two $10,000 nets and managed to catch more than 800 pounds of litter in the first few weeks. The full nets are emptied into special trucks and the refuse taken to a sorting center where it is separated into items that be recycled and items that cannot. With many more nets now installed, the system is expected to return its investment and become profitable in its first year.

Conscience

In *The Golden Age* (1889), Cuban author José Marti, wrote:

> There are men who can live contentedly even if they do live undignified lives. There are others who suffer as if in agony when they see people around them living without dignity. There must be a certain amount of dignity in the world. There must be a certain amount of light. When there are many undignified men, there are always others who have within them the dignity of many men. Those are the ones who rebel ferociously against those who rob nations of their freedom, which is robbing men of their dignity.

Those with the dignity of many men are beginning to act. In December 2017, the UN Environment Assembly adopted a global goal to stop the discharge of plastic to the sea.

In April 2018, just ahead of a Commonwealth Clean Oceans Alliance meeting in Vanuatu, British prime minister Theresa May announced her intent to ban single-use plastics, including straws and cotton swab handles. Calling plastic waste "one of the greatest environmental challenges facing the world," May said she would work with industry to develop alternatives. An estimated 8.5 billion plastic straws are tossed out in the UK every year, and £61.4 million from the public purse was pledged to fight the rising tide.

At that Commonwealth meeting led by Caribbean-born secretary general Patricia Scotland, fifty-two countries pledged to ban microbeads in rinse-off cosmetics and personal care products and to cut plastic bag use by 2021. On June 5, Indian prime minister Narendra Modi announced his intent to eliminate all single-use plastic in the country by 2022. With a fast-growing economy and population of 1.3 billion, India struggles to manage its vast waste stream and is a significant contributor to global ocean plastic. "Let us all join together to beat plastic pollution and make this planet a better place to live," Modi said.

On July 6, 2018, Chile's Constitutional Court ratified a bill that bans retail use of plastic bags across the country. The court ruled against an appeal of the law that had been filed by the plastics industry. Also in July, Seattle became the first US city to ban plastic straws and utensils in bars and restaurants. Roughly five thousand licensees are being told they must now switch to paper or compostable plastics. A similar ban that was proposed for Hawai'i was defeated by opposition from industry. Other proposed bans are being debated in San Francisco, New York, and Washington, DC, among other places.

In the same month, Disney, Starbucks, Red Lobster, and Lindblad Expeditions all announced they were getting rid of single-use plastics. McDonald's is also planning to phase out plastic straws at their UK and Ireland locations, coinciding with UK and EU proposals to cut single-use plastic. Family-owned Bacardi Ltd., the world's largest spirits producer with more than two hundred brands and labels, intends to cut its usage over the next two years by a billion straws.

The Aquarium Conservation Partnership (ACP), comprising twenty-two aquariums in seventeen US states, have gotten Alaska Airlines, American Airlines, United Airlines, the Chicago White Sox, and Dignity Health hospitals to stop offering plastic straws. It hopes to commit another five hundred companies by Earth Day, April 20, 2019. ACP is also partnering with the UN and European Commission to create a global coalition of two hundred aquariums to campaign against plastic.

In September 2018, Danish brewer Carlsberg became the first beer producer to ditch plastic multipack rings that hold beer and other cans together for holders made of recyclable glue.

In October, the European Parliament voted 571 to 53 to slash single-use plastic across the continent, beginning in 2021. The rules would also mandate that EU countries collect and recycle 90 percent of plastic bottles by 2025. Plastic producers would be on the hook for most of the expense of waste management and cleanup efforts.

The Ocean Cleanup

In September 2018, a six-year-old campaign to rid the world's oceans of plastic trash marked a milestone as a giant floating trash collector departed Alameda Shipyards in California bound for the Great Pacific Garbage Patch.

Did the engineers in the Netherlands invent the first feasible method for extracting large amounts of plastic debris from the sea? Over the course of the next year, the device will undergo the ultimate tests and face some tough questions: Can technology prevail over nature, or will the open Pacific tear it to shreds, turning the cleaner itself into plastic trash? And, even if the rigors of the ocean environment do not devour the device, will it attract marine animals, such as dolphins and turtles, and fatally entangle them?

The project is the creation of Boyan Slat, a twenty-four-year-old Dutch college dropout who raised more than $30 million to assemble a talented engineering team and build the machine. After a holiday diving trip in Greece impressed him with the scope of the problem, he quit his aerospace engineering studies at Delft University of Technology and made The Ocean Cleanup his life's mission. With the success of an initial crowdsourcing campaign, he was able to attract a top-notch staff of sixty-five dedicated engineers and scientists, many with years of experience designing platforms for oil companies.

Slat's effort won accolades and a Champions of the Earth award from the United Nations, but he was also taken to task by some scientists for underestimating the potential risks to marine life. Miriam Goldstein, director of ocean policy at the Center for American Progress, says the cleaning device can mimic a fish-aggregating device used by fishermen to draw oceangoing fish to a central area where they can be easily caught. As fish congregate at the device, they could then attract marine animals and become entangled.

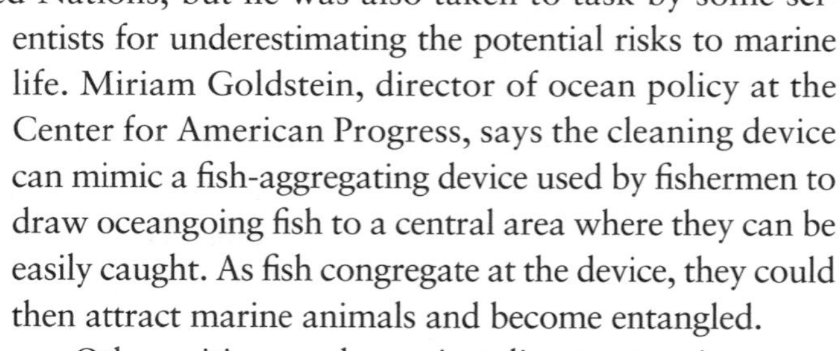

Boyan Slat

Other critics say the project diverts attention away from what is regarded as a more cost-effective, consequential way to save the oceans—by preventing plastic trash from flowing into it in the first place. "What's floating on the surface of the ocean gyres is only 3 percent of the plastic that enters the ocean every single year," says Eben Schwartz, marine debris program manager for the California Coastal Commission. "I understand why people are fascinated by this bright, shiny new object. But it's sort of a digital solution to an analog problem. The solution to plastic pollution entering our ocean starts on land."

Slat replied that it was his long-held view that prevention is the first step toward protecting the oceans but said, "I think it should be clear that humanity can do more than one thing at a time . . . the plastic in the ocean is not going to go away by itself. We see plastic dating back to the 1960s and 1970s, so I think it's obvious we have to do both. It's not a hopeful situation if the only thing you can do is not make it worse."

The system consists of a 600-meter-long floater that sits at the surface of the water and a tapered 3-meter-deep skirt attached below. The floater provides buoyancy to the system and prevents plastic from flowing over it, while the skirt stops debris from escaping underneath. After its first sea trials in San Francisco Bay, the floater, dubbed "Wilson" after a floating volleyball in the movie *Cast Away*, was dragged 350 miles out to sea for more trials in September 2018, before being sailed to the Great Pacific Garbage Patch. It arrived October 16 and was immediately deployed for a yearlong trial.

Slat's cleanup system has gone through several design changes. At first the boom, a high-density polyethylene pipe four feet in diameter and 1,969 feet long, was to be tethered to the ocean floor, but that scheme was scrapped in favor of a passive drifting system in which the floater naturally orients itself to the current and winds pull it into a U-shape while being slowed by a sea anchor set down in deeper, slower currents. Then that design was scrapped for a faster-drifting floater, propelled by wind and waves, that moves faster than the plastic, allowing the plastic to be captured. The free-floating floater, with only some solar-powered lights to ward off passing ships, cameras, and transmitters that send telemetry data to a satellite,

A floater, dubbed "Wilson," more than four football fields long, moves briskly along the surface, preventing large plastic from floating past, while a fixed skirt prevents smaller pieces from going under, without trapping fish.

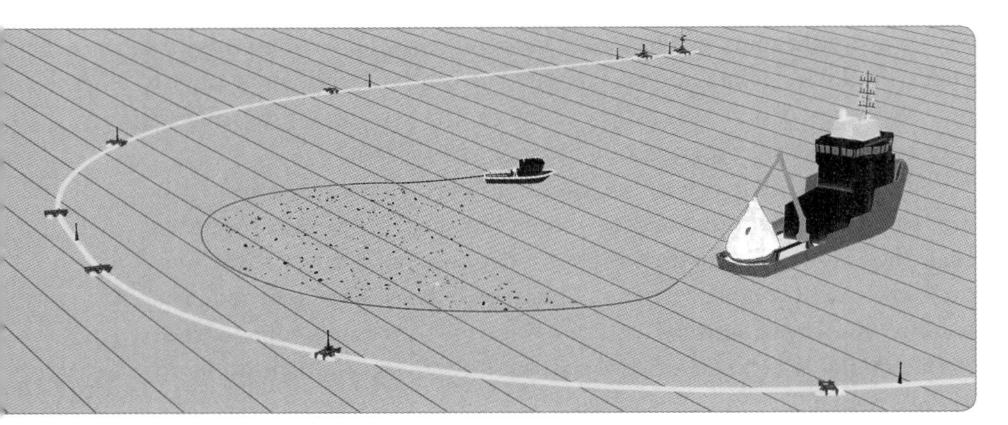

A smaller service vessel accompanies the mother ship to gather the plastic. These service ships arrive when sensors and cameras mounted on the boom tell them the location is ready to harvest.

moves slowly through the water gathering debris above 10 mm in size from up to ten feet below the surface.

Once it has trapped a sufficient volume of plastic, a service vessel arrives to collect it with a draw net and haul the net with a crane. The plastic is sorted and packed for delivery to recycling plants ashore.

At the start of 2019, the Pacific trial had already disproved Miriam Goldstein's theory. The floater held no interest for marine animals, and no fish or mammals were trapped in its skirt. As designers had expected, heavy seas exceeded the limits of the floater's tensile strength, and a section was sheared away by wave action, which ended the trial. Collection data also showed that the floater and skirt were not retaining captured plastic as well as they should. At this writing (January 2019), System 001 is in harbor at Hilo, Hawai'i, for analysis and retooling. By allowing waves to run through the system rather than toss it around or break over it, the Dutch engineers expect the next design to diffuse wind and wave energy to just 5 percent of the potential destructive force, allowing the floater to survive a one-hundred-year storm event.

Through this iterative process, The Ocean Cleanup will refine its approach until it achieves success.

Rotterdam could become the first city to pave its city streets, highways, and bike paths with plastic collected by The Ocean Cleanup. While the Lego Group itself is abandoning plastic, the city is experimenting with Lego-like polymer blocks in a project called Plastic Roads.

A spokesperson for the Plastic Roads developer, VolkerWessles, said, "Plastic offers all kinds of advantages compared to the current construction, both in road development and maintenance," including a greater endurance and ability to withstand more extreme temperatures ranging from -40 degrees C to 80 degrees C (-40 degrees F to 176 degrees F). Roads could also be laid in weeks, instead of months, because the blocks do not need to be built in place but are prefabricated, limiting the problems of logistics, service interruptions, and detours that usually occur during road construction.

The Ocean Cleanup should be able to harvest more than enough marine plastic to upcycle into paving Legos, and not just for Rotterdam. If the design process stays on schedule, by 2020, sixty additional systems will be at sea to gather more than fifteen thousand tons a year. The Ocean Cleanup estimates that with these simple systems—autonomous, energy neutral, and scalable—it can remove 50 percent of the Great Pacific Garbage Patch within five years and 90 percent of all ocean plastic by 2040. If you find this as inspiring a story as I do, please consider sending a donation to The Ocean Cleanup North Pacific Foundation.

The Final Word

Our plastics problem will get much worse unless we act, but like climate change, nuclear energy and weapons, or the extinction of species, we humans are slow to respond to these enormous changes we are causing to the natural world. Our powers, our numbers, and the complexity of our societies lie so far outside the historical experience for both ourselves and our world that we have a hard time perceiving the dangers we now face.

The science of neurobiology is only just coming to understand that human intelligence evolved because of an improbable adaptation our ancestors made at the dawn of consciousness that permits us to deny unpleasant realities. Ironically, it is that very adaptation that is responsible for our inability to recognize our severe state of ecological overshoot and so may cause our own extinction if we cannot react soon enough.

The only way to deal with the hard arithmetic of plastic pollution, or any of the other critical problems of this era, and to reduce the likelihood of our own annihilation is to override our genetic tendency to deny reality. In other words, we must alter one of the most fundamental features of the human character.

If our capacity to deny or ignore hard truths is our greatest liability, then our willingness to take up challenges and find ways to succeed is our greatest hope.

I hope that everyone who has stayed with me from the start of this book and gotten to this point has seen what I have seen and cannot deny that something must be done, quickly and urgently. It is also my hope that we can now agree we have solutions, and the only thing stopping us is us.

Bibliography

Abraham, Martin A. ed. *Sustainability Science and Engineering: Defining Principles*. Amsterdam: Elsevier Science, 2006.

Anastas, Paul T., Paul H. Bickart, and Mary M. Kirchhoff. *Designing Safer Polymers*. New York: Wiley-Interscience, 2000.

Anastas, Paul T., and John C. Warner. *Green Chemistry: Theory and Practice*. Oxford: Oxford University Press, 1998.

Campos, Cristian, ed. *Plastic*. New York: Collins Design, 2008.

Crawford, Christopher Blair, and Brian Quinn, eds. *Microplastic Pollutants*. Amsterdam: Elsevier Science, 2017.

Engineering Dictionaries. *Dictionary of Plastic Engineering*. Engineering Dictionaries, 2018. Kindle.

Fix, Alexandra. *Plastic*. Portsmouth, NH: Heinemann Educational Books, 2007.

Freinkel, Susan. *Plastic: A Toxic Love Story*. New York: Houghton Mifflin Harcourt, 2011.

Hall, Charles A. S., and Kent Klitgaard, eds. *Energy and the Wealth of Nations: An Introduction to Biophysical Economics*. 2nd ed. Cham, Switzerland: Springer Nature, 2018.

Hossain, Sahadat, Sadik Khan, and Golam Kibria. *Sustainable Slope Stabilisation Using Recycled Plastic Pins*. Boca Raton, FL: CRC Press, 2017.

Junger, Sebastian. *Tribe: On Homecoming and Belonging*. New York: Twelve, 2016.

Lundquist, Lars, Yves Leterrier, Paul Sunderland, and Jan-Anders Månson, eds. *Life Cycle Engineering of Plastics*. Oxford: Elsevier, 2000.

Mann, Charles C. *The Wizard and the Prophet: Two Remarkable Scientists and Their Dueling Visions to Shape Tomorrow's World*. New York: Knopf, 2018.

McCallum, Will. *How to Give Up Plastic: A Guide to Changing the World, One Plastic Bottle at a Time*. New York: Harper Design, 2018.

McCrum, N. G., C. P. Buckley, and C. B. Bucknall. *Principles of Polymer Engineering*. Oxford: Oxford University Press, 1988.

McKeen, Laurence W. *The Effect of UV Light and Weather on Plastics and Elastomers*. Norwich, NY: William Thomas, 2018.

Miller, Lorayne. *Survive! Plastic Bottle Wonder Hacks: A Prepper's Guide*. Self-published, Amazon Digital Services, 2018. Kindle.

Patrick, Katie. *Detrash Your Life in 90 Days: Your Complete Guide to the Art of Zero Waste Living*. Hello World Labs, 2016. Kindle.

Scheidt, Camille. *Unraveling Threads: How to Have a Sustainable Wardrobe in the Age of Plastic Fabric*. Self-published, Amazon Digital Services, 2018. Kindle.

Terry, Beth. *Plastic Free: How I Kicked the Plastic Habit and How You Can Too*. New York: Skyhorse, 2012.

Thomas, Sabu, Ajay Vasudeo Rane, Krishnan Kanny, Abitha V. K., and Martin George Thomas, eds. *Recycling of Polyethylene Terephthalate Bottles*. Oxford: William Andrew, 2018.

Tolinski, Michael. *Additives for Polyolefins*. Oxford: William Andrew, 2009.

Tolinski, Michael. *Plastics and Sustainability: Towards a Peaceful Coexistence between Bio-based and Fossil Fuel-based Plastics*. Hoboken, NJ: Wiley-Scrivener, 2011.

Yu, Long, ed. *Biodegradable Polymer Blends and Composites from Renewable Resources*. Hoboken, NJ: Wiley, 2009.

Wardman, Geordie. *Our Plastic Legacy: How to Quit Plastic, Want Less, and Live Green Daily*. Self-published, Amazon Digital Services, 2017. Kindle.

Zia, Khalid Mahmood, Mohammad Zuber, and Muhammad Ali, eds. *Algae Based Polymers, Blends, and Composites: Chemistry, Biotechnology and Materials Science*. Amsterdam: Elsevier, 2017.

Index

References for sidebars appear in *italics*.

About the Author

Albert Bates is a former attorney, paramedic, aid worker, communard, natural builder, educator, and the author of eighteen books on climate, history, and ecology, including *BURN: Using Fire to Cool the Earth* (2019), *The Paris Agreement* (2015), *The Biochar Solution: Carbon Farming and Climate Change* (2010), and *The Post-Petroleum Survival Guide and Cookbook* (2006). His book *Climate in Crisis* (1990) was among the first to call attention to potential for a runaway greenhouse effect in the twenty-first century. He is the founder and director of the Global Village Institute for Appropriate Technology (gvix.org), a nonprofit scientific research, development, and demonstration organization with projects on six continents, an ambassador for the Global Ecovillage Network, and an advisor to many organizations, foundations, and governments now applying regenerative design to reverse climate change.

For more books that inspire readers to create a healthy, sustainable planet for future generations, visit **BookPubCo.com**

Automating Hydroponics
Cerreto Rossouw
978-1-57067-366-5
$14.95

Growing Urban Orchards
Susan Poizner
978-1-57067-352-8
$19.95

Going Off the Grid
Gary Collins, MS
978-1-57067-354-2
$14.95

Tapping into Water
Paul Sawyers
978-1-57067-357-3
$15.95

Water Storage
Art Ludwig
978-0-96434-336-8
$19.95

How to Start a Worm Bin
Henry Owen
978-1-57067-349-8
$9.95

Purchase these titles from your favorite book source or buy them directly from:
Book Publishing Company • PO Box 99 • Summertown, TN 38483 • 1-888-260-8458

Free shipping and handling on all orders